Lecture Notes in Computer Science 11629

Commenced Publication in 1973
Founding and Former Series Editors:
Gerhard Goos, Juris Hartmanis, and Jan van Leeuwen

More information about this series at http://www.springer.com/series/7407

Esther Guerra · Fernando Orejas (Eds.)

Graph Transformation

12th International Conference, ICGT 2019
Held as Part of STAF 2019
Eindhoven, The Netherlands, July 15–16, 2019
Proceedings

 Springer

Editors
Esther Guerra 🆔
Universidad Autónoma de Madrid
Madrid, Spain

Fernando Orejas
Universitat Politècnica de Catalunya
Barcelona, Spain

ISSN 0302-9743   ISSN 1611-3349 (electronic)
Lecture Notes in Computer Science
ISBN 978-3-030-23610-6   ISBN 978-3-030-23611-3 (eBook)
https://doi.org/10.1007/978-3-030-23611-3

LNCS Sublibrary: SL1 – Theoretical Computer Science and General Issues

This Springer imprint is published by the registered company Springer Nature Switzerland AG
The registered company address is: Gewerbestrasse 11, 6330 Cham, Switzerland

Preface

This volume contains the proceedings of ICGT 2019, the 12th International Conference on Graph Transformation held during July 15–16, 2019 in Eindhoven, The Netherlands. ICGT 2019 was affiliated with STAF (Software Technologies: Applications and Foundations), a federation of leading conferences on software technologies. ICGT 2019 took place under the auspices of the European Association of Theoretical Computer Science (EATCS), the European Association of Software Science and Technology (EASST), and the IFIP Working Group 1.3, Foundations of Systems Specification.

The aim of the ICGT series is to bring together researchers from different areas interested in all aspects of graph transformation. Graph structures are used almost everywhere when representing or modelling data and systems, not only in computer science, but also in the natural sciences and in engineering. Graph transformation and graph grammars are the fundamental modelling paradigms for describing, formalizing, and analyzing graphs that change over time when modelling, e.g., dynamic data structures, systems, or models. The conference series promotes the cross-fertilizing exchange of novel ideas, new results, and experiences in this context among researchers and students from different communities.

ICGT 2019 continued the series of conferences previously held in Barcelona (Spain) in 2002, Rome (Italy) in 2004, Natal (Brazil) in 2006, Leicester (UK) in 2008, Enschede (The Netherlands) in 2010, Bremen (Germany) in 2012, York (UK) in 2014, L'Aquila (Italy) in 2015, Vienna (Austria) in 2016, Marburg (Germany) in 2017, and Toulouse (France) in 2018, following a series of six International Workshops on Graph Grammars and Their Application to Computer Science from 1978 to 1998 in Europe and in the USA.

This year, the conference solicited research papers describing new unpublished contributions in the theory and applications of graph transformation as well as tool presentation papers that demonstrate main new features and functionalities of graph-based tools. All papers were reviewed thoroughly by at least three Program Committee members and additional reviewers. We received 22 submissions, and the Program Committee selected 12 research papers and one tool presentation paper for publication in these proceedings, after careful reviewing and extensive discussions. The topics of the accepted papers range over a wide spectrum, including theoretical approaches to graph transformation, logic and verification for graph transformation, model transformation, as well as the application of graph transformation in some areas. In addition to these paper presentations, the conference program included an invited talk, given by Marieke Huisman (University of Twente, The Netherlands).

We would like to thank all who contributed to the success of ICGT 2019, the invited speaker Marieke Huisman, the authors of all submitted papers, as well as the members of the Program Committee and the additional reviewers for their valuable contributions to the selection process. We are grateful to Reiko Heckel, the chair of the Steering

Committee of ICGT for his valuable suggestions; to Mark van den Brand and Loek Cleophas, the general chair and the local chair, respectively, of STAF 2019; and to the STAF federation of conferences for hosting ICGT 2019. We would also like to thank EasyChair for providing support for the review process.

May 2019   Esther Guerra
                                                                Fernando Orejas

Organization

Steering Committee

Paolo Bottoni	Sapienza University of Rome, Italy
Andrea Corradini	University of Pisa, Italy
Gregor Engels	University of Paderborn, Germany
Holger Giese	Hasso Plattner Institute at the University of Potsdam, Germany
Reiko Heckel (Chair)	University of Leicester, UK
Dirk Janssens	University of Antwerp, Belgium
Barbara König	University of Duisburg-Essen, Germany
Hans-Jörg Kreowski	University of Bremen, Germany
Leen Lambers	Hasso Plattner Institute at the University of Potsdam, Germany
Ugo Montanari	University of Pisa, Italy
Mohamed Mosbah	LaBRI, University of Bordeaux, France
Manfred Nagl	RWTH Aachen, Germany
Fernando Orejas	Technical University of Catalonia, Spain
Francesco Parisi-Presicce	Sapienza University of Rome, Italy
John Pfaltz	University of Virginia, Charlottesville, USA
Detlef Plump	University of York, UK
Arend Rensink	University of Twente, The Netherlands
Leila Ribeiro	University Federal do Rio Grande do Sul, Brazil
Grzegorz Rozenberg	University of Leiden, The Netherlands
Andy Schürr	Technical University of Darmstadt, Germany
Gabriele Taentzer	University of Marburg, Germany
Jens Weber	University of Victoria, Canada
Bernhard Westfechtel	University of Bayreuth, Germany

Program Committee

Anthony Anjorin	University of Paderborn, Germany
Paolo Bottoni	Sapienza University of Rome, Italy
Andrea Corradini	University of Pisa, Italy
Juan De Lara	Autonomous University of Madrid, Spain
Juergen Dingel	Queen's University, Canada
Maribel Fernández	King's College London, UK
Holger Giese	Hasso Plattner Institute at the University of Potsdam, Germany
Esther Guerra (Co-chair)	Autonomous University of Madrid, Spain
Reiko Heckel	University of Leicester, UK
Barbara König	University of Duisburg-Essen, Germany

Harald König	FHDW Hannover, Germany
Leen Lambers	Hasso Plattner Institute at the University of Potsdam, Germany
Yngve Lamo	Western Norway University of Applied Sciences, Norway
Fernando Orejas (Co-chair)	Technical University of Catalonia, Spain
Detlef Plump	University of York, UK
Christopher M. Poskitt	Singapore University of Technology and Design, Singapore
Arend Rensink	University of Twente, The Netherlands
Leila Ribeiro	Federal University of Rio Grande do Sul, Brazil
Andy Schürr	Technical University of Darmstadt, Germany
Daniel Strüber	University of Koblenz and Landau, Germany
Gabriele Taentzer	Philipps University of Marburg, Germany
Jens Weber	University of Victoria, Canada

Additional Reviewers

Atkinson, Timothy
Heindel, Tobias
Kosiol, Jens
Lochau, Malte
Nolte, Dennis
Schneider, Sven
Tomaszek, Stefan
Wulandari, Gia

Contents

Transformation Rules Construction and Matching

Theory

Introducing Symmetry to Graph Rewriting Systems with Process Abstraction

Taichi Tomioka[✉], Yutaro Tsunekawa, and Kazunori Ueda

Waseda University, Tokyo, Japan
{tomioka,tsunekawa,ueda}@ueda.info.waseda.ac.jp

Abstract. Symmetry reduction in model checking is a technique for reducing state spaces by exploiting the inherent symmetry of models, i.e., the interchangeability of their subcomponents. Model abstraction, which abstracts away the details of models, often strengthens the symmetry of the models. Graph rewriting systems allow us to express models in such a way that inherent symmetry manifests itself with graph isomorphism of states. In graph rewriting, the synergistic effect of symmetry reduction and model abstraction is obtained under graph isomorphism. This paper proposes a method for abstracting programs described in a hierarchical graph rewriting language LMNtal. The method automatically finds and abstracts away subgraphs of a graph rewriting system that are irrelevant to the results of model checking. The whole framework is developed within the syntax and the formal semantics of the modeling language LMNtal without introducing new domains or languages. We show that the proposed abstraction method combined with symmetry reduction reduces state spaces while preserving the soundness of model checking. We implemented the method on SLIM, an implementation of LMNtal with an LTL model checker, tested it with various concurrent algorithms, and confirmed that it automatically reduces the number of states by successfully extracting the symmetry of models.

Keywords: LMNtal · Graph rewriting systems · Model checking · Symmetry reduction · Abstraction

1 Introduction

Symmetry reduction [3] in model checking is one of the key techniques to tackle the space explosion problem. Symmetry reduction reduces a state space by regarding equivalent states (i.e., states that need not be distinguished from each other) in the state space as a single state using the symmetry of a model. The technique should be applicable to diverse models; for instance, analysis of algorithms for board games and of distributed algorithms over a network of processes may well exploit the symmetry of the board and of the network topology, respectively. In general, symmetry of a model is obtained by program analysis [5] or given by user specification [17].

© Springer Nature Switzerland AG 2019
E. Guerra and F. Orejas (Eds.): ICGT 2019, LNCS 11629, pp. 3–20, 2019.
https://doi.org/10.1007/978-3-030-23611-3_1

There are many model description languages for model checking, ranging from concurrent imperative languages like Promela [12] to rewriting systems. Of these, graph rewriting systems are a highly general and expressive modeling framework in the sense that they allow us to represent both dynamically evolving *processes structures* formed by processes, channels and messages and dynamically evolving *data structures* formed by constructors and pointers in a uniform manner. They are general also in the sense that they subsume multiset rewriting which is the basis of the Gamma model, Chemical Abstract Machines and also Petri Nets. Furthermore, they are highly amenable to symmetry reduction because graph rewriting systems allow us to express individual states as graphs in such a way that various kinds of symmetry such as rotation and reflection manifest themselves. The symmetry can then be handled as graph isomorphism checking in graph rewriting systems, as pioneered by the model checker of the graph transformation tool GROOVE [9,18].

However, even when a model appears to be symmetric basically, the symmetry may not be recognized as graph isomorphism due to minor differences that break the symmetry. For example, consider the model checking of concurrent programs. Concurrent processes and channels are usually given their own IDs, either explicitly or implicitly. These IDs and other inessential details must be carefully eliminated to reveal the symmetry.

Some asymmetric states represented as graphs could possibly be made symmetric by abstracting them appropriately. This motivates us to consider eliminating inessential details of the models and make previously nonequivalent states equivalent. It may enable the symmetry reduction mechanism to reduce states even further. In our approach we will detail in this paper, which parts of a model are irrelevant to verification and can be abstracted away is automatically decided from a model described as a rewriting system and its verification conditions. The advantages of the technique are twofold. Firstly, space and time efficiency can be improved without working on complicated models manually. Secondly, it enables us to generate models which are simplified but yet detailed enough to validate given properties. For example, we can reuse the same model for different verification conditions and apply model abstraction only for the conditions that allow the use of the model's inherent symmetry. We use LMNtal [26][1], a language based on hierarchical graph rewriting, as a model description language. Hierarchical graphs of LMNtal consist of labeled nodes, links, and membranes that enclose subgraphs to form hierarchical structures. Many systems in the world have two kinds of structures that may co-exist: connectivity and hierarchy. LMNtal provides these two structuring mechanisms for the straightforward modeling of various systems and formalisms. Its expressive power has been demonstrated by the encoding of various formalisms ranging from the ambient calculus [24] to strong reduction of the λ-calculus [25]. Formalisms designed with a similar motivation include Milner's bigraphs [16]. SLIM [10][2], a full-fledged implementation of LMNtal, features state space construction and LTL model checking.

---

[1] LMNtal homepage: https://www.ueda.info.waseda.ac.jp/lmntal/.
[2] Available at https://github.com/lmntal/slim.

This paper contains three main contributions. First, we developed a method for *automatically* reducing the state space of a model by *static* model abstraction that preserves the soundness of graph-based model checking. Second, we showed the soundness of abstraction by reducing equivalence relations *induced by the abstraction* to equivalence relations *defined over LMNtal programs*, that is, without introducing yet another formalism or sublanguage. Third, we established a relationship between (i) SLIM's symmetry reduction based on the structural equivalence and reduction relation of LMNtal and (ii) symmetry reduction known as a standard formulation of model checking [4].

We implemented the proposed method on SLIM, made experiments with examples including concurrent algorithms, and confirmed that our method successfully reduces the state space.

The rest of this paper is organized as follows. Section 2 introduces related work. Section 3 introduces the hierarchical graph rewriting language LMNtal and its implementation SLIM. Section 4 explains how symmetry reduction is realized in LMNtal. Section 5 describes our method for model abstraction. Section 6 shows the results of experiments.

2   Related Work

Reduction of state space by abstracting graphs for graph rewriting systems has been studied before. Rensink et al. [19] proposed a method for abstracting graph rewriting systems. An abstract graph rewriting system is created automatically by abstracting graph generation rules and initial state graphs. Backes et al. [1] focused on the local relationships of graphs and abstract them into chunks of nodes called clusters. In these papers, the authors consider an abstract graph rewriting system with different semantics from the original and discuss an abstract program expressed in the abstract system. Our work differs from them in expressing abstract programs in the original language instead of giving additional framework. Also, it is worth noting that we have developed a framework of abstraction and its proof technique for a graph rewriting language whose syntax and semantics are defined in a structural manner, i.e., in a standard style of concurrent languages.

The abstraction of graphs proposed in this paper enhances the effect of symmetry reduction as will be described in Sect. 5. Sistla et al. [23] proposed a symmetry reduction method that exploits *process symmetry* and *state symmetry* of a model. Process symmetry is a relation between two global states that can be considered equivalent. State symmetry exploits the symmetry of a single state composed of possibly many processes with the same state. Since the graph rewriting systems we are working on express the state of a program as a graph, the two kinds of symmetry are expressed as graph isomorphism (that includes graph automorphism) comprehensively.

There is a lot of work on symmetry reduction for Petri Nets, whose survey can be found in [14]. The techniques can be divided into (i) reduction based on the isomorphism of the states of Place/Transition Nets [21] and (ii) reduction

(process) $P ::=$ **0**	(null)
$\mid p(X_1, \ldots, X_m) \quad (m \geq 0)$	(atom)
$\mid P, P$	(molecule)
$\mid \{P\}$	(cell)
$\mid T :\text{-} T$	(rule)
(process template) $T ::=$ **0**	(null)
$\mid p(X_1, \ldots, X_m) \quad (m \geq 0)$	(atom)
$\mid T, T$	(molecule)
$\mid \{T\}$	(cell)
$\mid T :\text{-} T$	(rule)
$\mid \$p[X_1, \ldots, X_m \mid *X] \quad (m \geq 0)$	(process context)

Fig. 1. Syntax of LMNtal (constructs not relevant to this paper are omitted.)

based on the data values of high-level Petri Nets such as CPN [13]. In LMNtal, on the other hand, the symmetry of states is handled uniformly by structural congruence which will be described in Sect. 3. Since graphs (with labeled nodes) are an expressive structure that subsumes lists, trees and scalar values, and structural congruence has been built into the language, LMNtal allows one to describe symmetric models without specifying symmetry explicitly.

Sistla et al. [22] and Emerson et al. [6] discussed methods to reduce asymmetric models by symmetry reduction. Our study is similar to theirs in that both strengthen the symmetry of models and reduce their state space while preserving soundness. The difference is that we give a concrete framework and implementation for a full-fledged language and that the method is applicable to a wide variety of models.

Symmetry of graph-based models has been studied in contexts other than model checking. For instance, Feret proposed a method for analyzing properties of models described in Kappa [8] but their research goals and the underlying formalisms are quite different from ours.

3   LMNtal: Graph Rewriting Language

This section describes a hierarchical graph rewriting language LMNtal and its implementation SLIM. Readers are referred to [26] for further details including design principles and relation to other computational models.

3.1   Syntax of LMNtal

LMNtal is a programming and modeling language for describing graph rewriting systems composed of *hierarchical graphs* (which we will simply call *graphs* henceforth) and *rewrite rules*. Graphs handled by LMNtal are (connected or unconnected) undirected graphs with labeled nodes. The syntax of LMNtal is

```
% Initial state
phi(L1,R1), {+R1,+L2},
phi(L2,R2), {+R2,+L3},
phi(L3,R3), {+R3,+L4},
phi(L4,R4), {+R4,+L5},
phi(L5,R5), {+R5,+L1},
% grab a left fork
({+X,+L}, phi(L,R) :- {-X,+L}, phi(L,R)),
% grab a right fork
({-X,+L}, phi(L,R), {+R,+Y} :- {-X,+L}, phi(L,R), {+R,-Y}),
% release forks
({-X,+L}, phi(L,R), {+R,-Y} :- {+X,+L}, phi(L,R), {+R,+Y}).
```

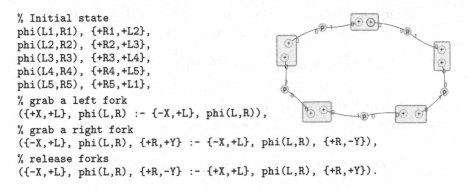

Fig. 2. A dining philosophers model (left) and its visualization by LaViT (right)

shown in Fig. 1. We call LMNtal graphs *processes* since the language was initially designed to model concurrency. The **0** stands for an empty process. A
graph node is called an *atom* and may have edges called *links*. Unlike standard
graphs found in the literature of graph theory and graph rewriting [20], *the links
connected to an atom are totally ordered* and each atom has a fixed arity (number of links). Atom names start with lowercase letters, and link names start with
capital letters. Each link name can occur at most twice in a process because it
expresses an endpoint of an edge. We call a link whose name occurs once in a
process a *free link*, and a link whose name occurs twice in a process a *local link*.
A *molecule* stands for parallel composition of processes. A *membrane* is provided
to introduce hierarchical structure to processes. A *cell* refers to processes contained in a membrane, including the membrane itself. A cell could be considered
as a composite graph node. A rewrite rule is composed of two process templates.

Process templates are the same as processes except that they may contain
process contexts. A *process context* matches non-rule processes and works as a
wildcard in rewriting. Process contexts are to be specified within membranes.
Arguments of a process context specify what free links may or must occur in a
matched process; the details are omitted and can be found in [26].

There is a special binary atom = called a *connector*. An atom $= (X, Y)$ fuses
link X and link Y and can be written as $X = Y$.

To give the connection between the above syntax and the graph structure
it represents[3], Fig. 2 shows an example LMNtal program modeling the dining
philosophers problem, whose initial state forms a ring structure formed by atoms
(representing philosophers) and membranes (representing shared forks), and its
visualization by LaViT (LMNtal Visual Tool), a publicly available integrated
development environment of LMNtal. The unary atoms '+'(X) and '−'(X)
(representing *"the fork is not in use"* and *"the fork is in use"*, respectively) are
written as prefix operators +X and −X, respectively.

---

[3] The details of this correspondence are beyond the scope of the present paper and
not described here.

(E1) $\mathbf{0}, P \equiv P$ (E2) $P, Q \equiv Q, P$ (E3) $P, (Q, R) \equiv (P, Q), R$

(E4) $P \equiv P[Y/X]$ if X is a local link of P

(E5) $\dfrac{P \equiv P'}{P, Q \equiv P', Q}$ (E6) $\dfrac{P \equiv P'}{\{P\} \equiv \{P'\}}$

(E7) $X = X \equiv \mathbf{0}$ (E8) $X = Y \equiv Y = X$

(E9) $X = Y, P \equiv P[Y/X]$ if P is an atom and X occurs free in P

(E10) $\{X = Y, P\} \equiv X = Y, \{P\}$ if exactly one of X and Y occurs free in P

Fig. 3. Structural congruence on LMNtal processes [26]

We denote by $P[Y/X]$ a process P whose links X are replaced with Y and by $P[Y \leftrightarrow X]$ a process P with all occurrences of X and Y replaced by each other. The set of all processes generated by the syntax rules is denoted as **Proc**. We also denote the set of all atoms as **Atom**, and the set of all links as **Link**. We may use metavariables $p, q, \ldots$ to indicate atoms, $X, Y, \ldots$ for links, $P, Q, \ldots$ for processes, and $T, U, \ldots$ for process templates.

Unlike standard data structures like lists and trees, graphs whose links may interconnect nodes in a arbitrary manner do not allow straightforward inductive arguments. How to formalize concepts like symmetry and abstraction over LMNtal's powerful hierarchical graph structure is therefore a challenging topic. However, terms generated by the LMNtal syntax have a tree structure, over which structural congruence to be discussed in Sect. 3.2 is defined to allow an interpretation of them as graphs or network of nodes (which in turn allow an interpretation as processes, messages, data constructors etc.). Accordingly, we use *paths* over a syntax tree to refer to elements of terms. Given a term P, we consider a partial function $(\!|P|\!) : \mathbb{N}_{>0}{}^* \to \{,, \{\}, :-, \mathbf{0}\} \cup \mathbf{Atom} \cup \mathbf{Link}$ to refer to individual occurrences of symbols. The symbol $\bullet$ expresses an empty sequence of positive integers. $(\!|P|\!)$ is defined inductively as follows:

$$(\!|\mathbf{0}|\!)(\pi) := \mathbf{0} \quad (\pi = \bullet) \qquad (\!|p(X_1, \ldots, X_n)|\!)(\pi) := \begin{cases} p & (\pi = \bullet) \\ X_i & (\pi = i) \end{cases}$$

$$(\!|P, Q|\!)(\pi) := \begin{cases} , & (\pi = \bullet) \\ (\!|P|\!)(\rho) & (\pi = 1\rho) \\ (\!|Q|\!)(\rho) & (\pi = 2\rho) \end{cases} \qquad (\!|\{P\}|\!)(\pi) := \begin{cases} \{\} & (\pi = \bullet) \\ (\!|P|\!)(\rho) & (\pi = 1\rho) \end{cases}$$

$$(\!|T :- U|\!)(\pi) := T :- U \quad (\pi = \bullet)$$

As suggested by the above definition, we do not consider the internal structure of rewrite rules. It is not difficult to see that there is one-to-one correspondence between P and $(\!|P|\!)$. Note also that π is uniquely determined for all elements of a term. Thus, we can regard π as the *identifier* of a syntactic element $(\!|P|\!)(\pi)$. Given terms P, Q and a path α, $(\!|P|\!)[\alpha \mapsto Q]$ denotes a term P with the subtree rooted at α replaced by Q.

$$\text{(R1)} \quad \frac{P \longrightarrow P'}{P, Q \longrightarrow P', Q} \qquad \text{(R2)} \quad \frac{P \longrightarrow P'}{\{P\} \longrightarrow \{P'\}} \qquad \text{(R3)} \quad \frac{Q \equiv P \quad P \longrightarrow P' \quad P' \equiv Q}{Q \longrightarrow Q'}$$

$$\text{(R4)} \quad \{X = Y, P\} \longrightarrow X = Y, \{P\} \quad \text{if } X \text{ and } Y \text{ occur free in } \{X = Y, P\}$$

$$\text{(R5)} \quad X = Y, \{P\} \longrightarrow \{X = Y, P\} \quad \text{if } X \text{ and } Y \text{ occur free in } P$$

$$\text{(R6)} \quad T\theta, (T :\text{-} U) \longrightarrow U\theta, (T :\text{-} U)$$

Fig. 4. Reduction relation on LMNtal processes [26]

3.2 Relations Between Processes

The syntax we introduced is not sufficient to represent the graph structure of LMNtal processes. There is a pair of processes which are regarded as isomorphic but are not (yet) equal terms. For example, the names of local links and the order of atoms are insignificant when interpreting processes as graphs with unlabeled edges. To absorb these syntactic differences, LMNtal provides *structural congruence*, which is an equivalence relation on processes, as defined in Fig. 3.

Next, we explain computation of LMNtal processes. Rewriting of processes is defined by a reduction relation ($\longrightarrow$) shown in Fig. 4. There are six rules of which the most fundamental one is (R6). The θ in (R6) is a mapping from process contexts in T to concrete processes. This means that the θ is identity if the rule $T :\text{-} U$ contains no process contexts. If a rewrite rule is applicable to multiple subprocesses of a process, the language does not define which is rewritten, that is, the rewriting can be nondeterministic.

For example, consider the LMNtal process with three rewrite rules:

```
p(P), waiting(P), q(Q), waiting(Q), semaphore,
   (waiting(X) :- requesting(X)),
   (requesting(X), semaphore :- processing(X)),
   (processing(X) :- waiting(X), semaphore).
```

The first rewrite rule is applicable to two parts of the initial process. The state space generated by this program is shown in Fig. 5. There are other structurally congruent processes for each state, but here we have shown just one of those processes. Rewrite rules in processes are omitted because all processes have the same set of rules.

We normally avoid redundancy in discussion by equating structurally congruent processes implicitly. However, this makes it difficult to discuss the effect of structural congruence closely, which is exactly the topic of the present paper. Consequently, unless stated otherwise, we deal with structural congruence explicitly via the relation $\equiv$.

3.3 Semantics of LMNtal

Computation in LMNtal proceeds by applying rewrite rules to processes. In this paper, we define the semantics $[\![P]\!]$ of a term P as its state space, i.e., the state transition graph generated by nondeterministic rewriting.

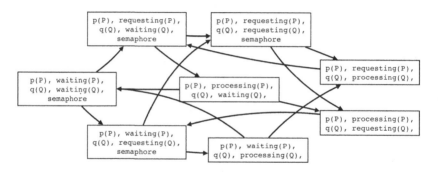

Fig. 5. State space of an LMNtal program (rules are omitted)

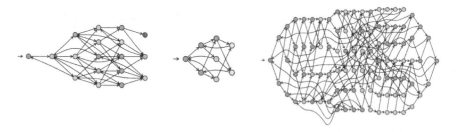

Fig. 6. State space diagrams of LMNtal programs generated by LaViT; Dining Philosophers with one deadlock state (left), the semaphore program (middle) and the Peterson algorithm with abstraction (right)

Definition 1. *Semantics $[\![P]\!]$ of an LMNtal process P is a state transition graph represented by a triplet S_P, R_P, P, where*

1. *S_P is the set of all states reachable from P,*
2. *R_P is a binary relation on S_P generated by LMNtal's reduction rules,*
3. *P is the initial state.*

In model checking, each state is augmented with a labeling function $L : S_P \rightarrow 2^{AP}$ that judges whether the state satisfies each atomic proposition in AP. An atomic proposition $ap \in AP$ here is represented by an LMNtal term T :- **0**. The left-hand side T works as a predicate on a process P and judges whether T matches P or its subprocess. In other words, if T :- **0** can rewrite P, P satisfies ap. By abuse of notation, we denote an augmented transition system (S_P, R_P, P, L) also as $[\![P]\!]$.

LaViT features a visualizer of $[\![P]\!]$. Figure 6 shows the state space of the dining philosophers program (Sect. 3.1), the semaphore program (Sect. 3.2) and the Peterson algorithm (Sect. 6) visualized by LaViT.

4 Symmetry Reduction

Symmetry reduction [15] in model checking reduces the state space of a model with the symmetry inherent in the model. We can obtain smaller state spaces by equating symmetric states. Symmetry reduction is achieved by defining equivalent classes of states that preserve the transition relation. State spaces reduced with such an equivalence is sound in CTL* (and hence LTL) model checking [7].

Although our process abstraction method presented in Sect. 5 reveals the potential symmetry of a model, it is necessary to use symmetry reduction after the abstraction in order to reduce the state space of the model. LMNtal and its implementation SLIM have long exploited the symmetry of models implicitly, but the formal argument of symmetry reduction has not been made so far. In this section, we show how state spaces in LMNtal can be reduced by structural congruence and symmetry reduction. We first redefine structural congruence as a symmetric group of mapping between processes. Then we show that the group enjoys the soundness condition of symmetry reduction.

4.1   Symmetric Group for Structural Congruence

The standard theory of symmetry reduction employs the notion of the symmetric group of bijective mappings σ between states to define equivalence relations between states. On the other hand, the structural congruence of LMNtal is defined in a syntax-directed manner. In general, an equivalence relation has an underlying symmetric group generating the relation. We redefine the structural congruence as a symmetric group in order to discuss symmetry reduction in LMNtal in a clearer way. That is, we define a symmetric group E of mappings $\sigma : \mathbf{Proc} \to \mathbf{Proc}$ and prove that an equivalence relation on processes given by E is equivalent to the structural congruence. This is not straightforward because (E5) and (E6) of the structural congruence rules are defined inductively on the LMNtal syntax. They make the definition of E somewhat tricky. Here we give mappings corresponding to the structural congruence rules except (E5) and (E6) as bases, and E is defined as the least fixed point of a generating function.

First, we define a family of fundamental mappings in Fig. 7. Each mapping except $\sigma_{1,n}$, $\sigma_{4,X \leftrightarrow Y}$, $\sigma_{7,X}$, and $\sigma_{9,X,i}$ maps the LHS to the RHS and the RHS to the LHS of structural congruence rules. Note that if $\sigma_{1,1}$ were defined just like $\mathbf{0}, P \leftrightarrow P$, $\sigma_{1,1}$ would not be bijective because $\sigma_{1,1}$ maps $\mathbf{0}, \mathbf{0}, P$ and P equally into $\mathbf{0}, P$. $\sigma_{1,n}$ defined in Fig. 7 maps multiple $\mathbf{0}$s into different processes or themselves. $\sigma_{4,X \leftrightarrow Y}$ changes a local link name with a new link name which does not occur in a source process. $\sigma_{9,X,i}$ corresponds to (E9) which is both an introduction rule and an elimination rule of a connector. X stands for the name of a link introduced or eliminated, and i stands for the position of the link. The inverses of $\sigma_{1,n}$, $\sigma_{4,X \leftrightarrow Y}$, $\sigma_{7,X}$, and $\sigma_{9,X,i}$ are themselves like other mappings.

Let E_1 be $\{\sigma_{1,n} \mid n \in \mathbb{N}\}$, E_2 be $\{\sigma_{4,X \leftrightarrow Y} \mid X, Y \in \mathbf{Link}\}$, E_3 be $\{\sigma_{7,X} \mid X \in \mathbf{Link}\}$, E_4 be $\{\sigma_{9,X,i} \mid X \in \mathbf{Link}, i \in \mathbb{N}_{>0}\}$, and E_5 be $\{\sigma_2, \sigma_3, \sigma_8, \sigma_{10}\}$, then a generating function F is defined as

$$F(A) := \bigcup_{i=1}^{5} E_i \cup \{\sigma \circ \tau \mid \sigma, \tau \in A\} \cup \{\mu_5 \sigma \mid \sigma \in A\} \cup \{\mu_6 \sigma \mid \sigma \in A\}.$$

$$\sigma_{1,n}((\!|s|\!)) := \begin{cases} (\!|s|\!) & (s = \underbrace{\mathbf{0},\ldots,\mathbf{0}}_{k}, P \wedge 2n \leq k) \\ (\!|\underbrace{\mathbf{0},\ldots,\mathbf{0}}_{k-n}, P|\!) & (s = \underbrace{\mathbf{0},\ldots,\mathbf{0}}_{k}, P \wedge n \leq k < 2n) \\ (\!|\underbrace{\mathbf{0},\ldots,\mathbf{0}}_{k+n}, P|\!) & (s = \underbrace{\mathbf{0},\ldots,\mathbf{0}}_{k}, P \wedge 0 \leq k < n) \end{cases}$$

$$\sigma_2((\!|s|\!)) := \begin{cases} (\!|Q,P|\!) & (s = P,Q) \\ (\!|s|\!) & \text{(otherwise)} \end{cases} \qquad \sigma_3((\!|s|\!)) := \begin{cases} (\!|P,(Q,R)|\!) & (s = (P,Q),R) \\ (\!|(P,Q),R|\!) & (s = P,(Q,R)) \\ (\!|s|\!) & \text{(otherwise)} \end{cases}$$

$$\sigma_{4,X \leftrightarrow Y}((\!|s|\!)) := \begin{cases} (\!|s[X \leftrightarrow Y]|\!) & \text{(Either } X \text{ or } Y \text{ is local and the other is free in } s) \\ (\!|s|\!) & \text{(otherwise)} \end{cases}$$

$$\sigma_{7,X}((\!|s|\!)) := \begin{cases} (\!|X\texttt{=}X|\!) & (s = \mathbf{0}) \\ (\!|\mathbf{0}|\!) & (s = X\texttt{=}X) \\ (\!|s|\!) & \text{(otherwise)} \end{cases} \qquad \sigma_8((\!|s|\!)) := \begin{cases} (\!|Y\texttt{=}X|\!) & (s = X\texttt{=}Y) \\ (\!|s|\!) & \text{(otherwise)} \end{cases}$$

$$\sigma_{9,X,i}((\!|s|\!)) := \begin{cases} (\!|P[Y/X]|\!) & (s = X\texttt{=}Y, P \text{ and } P \text{ is an atom and } (\!|P|\!)(i) = X) \\ (\!|X\texttt{=}Y, P|\!)[2i \mapsto X] & (s = P \text{ and } P \text{ is an atom and } (\!|P|\!)(i) = Y) \\ (\!|s|\!) & \text{(otherwise)} \end{cases}$$

$$\sigma_{10}((\!|s|\!)) := \begin{cases} (\!|\{X\texttt{=}Y, P\}|\!) & (s = X\texttt{=}Y, \{P\} \text{ and exactly one of } X \text{ and } Y \text{ occurs free in } P) \\ (\!|X\texttt{=}Y, \{P\}|\!) & (s = \{X\texttt{=}Y, P\} \text{ and exactly one of } X \text{ and } Y \text{ occurs free in } P) \\ (\!|s|\!) & \text{(otherwise)} \end{cases}$$

Fig. 7. A family of fundamental mappings

where $\mu_5\sigma$ and $\mu_6\sigma$ are defined as

$$\mu_5\sigma((\!|s|\!))(\pi) := \begin{cases} \sigma((\!|P|\!))(\rho) & (s = P,Q \wedge \pi = 1\rho) \\ (\!|s|\!)(\pi) & \text{(otherwise)}, \end{cases}$$

$$\mu_6\sigma((\!|s|\!))(\pi) := \begin{cases} \sigma((\!|P|\!))(\rho) & (s = \{P\} \wedge \pi = 1\rho) \\ (\!|s|\!)(\pi) & \text{(otherwise)}. \end{cases}$$

The inverses of $\mu_5\sigma$ and $\mu_6\sigma$ are

$$(\mu_5\sigma)^{-1}((\!|s|\!))(\pi) = \begin{cases} \sigma^{-1}((\!|P|\!))(\rho) & (s = P,Q \wedge \pi = 1\rho) \\ (\!|s|\!)(\pi) & \text{(otherwise)}, \end{cases}$$

$$(\mu_6\sigma)^{-1}((\!|s|\!))(\pi) = \begin{cases} \sigma^{-1}((\!|P|\!))(\rho) & (s = \{P\} \wedge \pi = 1\rho) \\ (\!|s|\!)(\pi) & \text{(otherwise)}. \end{cases}$$

Now we construct a group E whose underlying set is the least fixed point of F and whose multiplication is function composition. An equivalence relation $\sim_E$ on **Proc** introduced by E is defined as follows:

$$P \sim_E Q \iff \exists \sigma \in E.\ (\!|Q|\!) = \sigma(\!|P|\!)$$

Then, we can show that $\equiv$ and $\sim_E$ are equivalent (proof in Appendix):

Theorem 1. *For all processes* $P, Q \in \mathbf{Proc}$, $P \equiv Q \iff P \sim_E Q$.

4.2 Soundness of Structural Congruence

Let $[s]_E$ denote $\{\sigma s \mid \sigma \in E\}$, the equivalence class of $s \in \mathbf{Proc}$ by a symmetric group E. We denote some representative element of $[s]_E$ as $rep([s]_E)$. Now we define $[\![P]\!]/{\equiv} = (S_P/{\equiv}, R_P/{\equiv}, P/{\equiv})$ as follows:

$$S_P/{\equiv} := \{rep([s]_E) \mid s \in S_P\}$$
$$R_P/{\equiv} := \{(rep([s]_E), rep([t]_E)) \mid (s, t) \in R_P\}$$
$$P/{\equiv} := rep([P]_E)$$

For $[\![P]\!]/{\equiv}$ to be sound with respect to model checking, it is sufficient that $\sigma \in E$ preserves the reduction relations. By the reduction rule (R3), we have

$$\forall s, s', t, t' \in \mathbf{Proc},\ s \longrightarrow t \wedge s \equiv s' \wedge t \equiv t' \implies s' \longrightarrow t'$$
$$\Leftrightarrow \quad \forall s, s', t, t' \in \mathbf{Proc}, (\exists \sigma, \tau \in E,\ s \longrightarrow t \wedge s' = \sigma s \wedge t' = \tau t) \implies s' \longrightarrow t'$$
$$\Leftrightarrow \quad \forall s, s', t, t' \in \mathbf{Proc}, \forall \sigma, \tau \in E, (s \longrightarrow t \wedge s' = \sigma s \wedge t' = \tau t) \implies s' \longrightarrow t'$$
$$\Leftrightarrow \quad \forall s, t \in \mathbf{Proc}, \forall \sigma, \tau \in E,\ s \longrightarrow t \implies \sigma s \longrightarrow \tau t.$$

Hence all mappings $\sigma \in E$ satisfy the soundness of the reduction in symmetry reduction: $\forall s, t,\ (s, t) \in R \implies (\sigma(s), \sigma(t)) \in R$. As a result, for all LTL formula ϕ, $[\![P]\!]/{\equiv} \models \phi \implies [\![P]\!] \models \phi$ holds by a discussion similar to symmetry reduction.

5 Process Abstraction

We propose a process abstraction method for introducing further symmetry to reduce state spaces. Our method eliminates some details of processes in order to make them equivalent under structural congruence. For example, in the semaphore program in Sect. 3.2, the choice of the atom names p and q is irrelevant in the sense that it does not affect the form of the state space; that is, the state transition graphs are isomorphic up to the choice of these atom names. This means that, unless the verification condition mentions those atom names explicitly, p and q are not the essence of this model. Now we consider a variant of the program which uses a special atom name # instead of p and q:

```
#(P), waiting(P), #(Q), waiting(Q), semaphore,
  (waiting(X) :- requesting(X)),
  (requesting(X), semaphore :- processing(X)),
  (processing(X) :- waiting(X), semaphore)
```

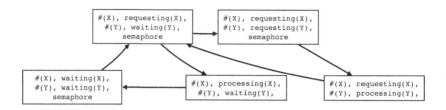

Fig. 8. Reduced state space of the semaphore program

The state space of this program (Fig. 8) is smaller than that of the previous program (Fig. 5). Indeed, the change of these atom names can be shown to be a graph homomorphism over the state space (Sect. 5.2), and we can still see in Fig. 8 that these two processes are not in the **processing** state simultaneously.

In this way, we may be able to obtain reduced state spaces by abstracting away some part of processes and thus enlarging equivalence classes based on the structural congruence relation. We will explain how to perform process abstraction while keeping the soundness of model checking.

5.1 UPE, Unused Process Elimination

First, we give some notations to explain our algorithm. Suppose $(\![P]\!)(\pi) = p$ and $(\![P]\!)(\pi i) = X_i\ (i = 1 \ldots n)$. We denote the arity of an atom π as $\mathsf{num}(\pi)$ and the atom connected to the ith argument as $\mathsf{atom}(\pi, i)$. The function $\mathsf{membrane}(\pi)$ returns the innermost membrane containing π, which can be obtained by finding the longest prefix π' of π satisfying $(\![P]\!)(\pi') = \{\}$. If π is not contained in any membrane, $\mathsf{membrane}(\pi)$ returns $\bullet$. Conversely, by $\mathsf{atoms}((\![P]\!), \pi_M)$ we denote a function that returns the set of all atom occurrences directly contained in a membrane π_M. If $\pi = \bullet$, it returns the set of all atom occurrences not included in any membrane. In addition, $\mathsf{atoms}((\![P]\!))$ returns the set of all atom occurrences of P, disregarding the membranes of P.

Algorithm 1 shows our algorithm UPE (Unused Process Elimination) for process abstraction. If an atom name p occurs in a rewrite rule, atoms with the name p may be rewritten by the rewrite rule. Let F be the set of atom names occurring in the rewrite rules in a process. For each atom in the process, UPE checks whether its name is contained in F and if not, marks it as removable. An atom whose name is not in F will not be a target of rewriting, so deleting the atom should make no difference to the behavior of a program. Because links between an atom to be deleted and another atom must be retained, the endpoint of a link connected to a deleted atom is terminated by a unary atom with a fresh name (for which we use # in this paper) not occurring in P. Free links crossing a membrane are terminated in the same way because arguments of a process context may match these free links (Fig. 9). When all atoms directly included in a membrane are deleted, UPE adds a special nullary atom in order to indicate that there were some atoms in the membrane (Fig. 10). Added atoms and deleted atoms are recorded in the set A and D for rewriting, respectively. Note that UPE

Algorithm 1. UPE, Unused Process Elimination

---

function UPE(P)

    $A := \emptyset$

    $D := \emptyset$

    $F :=$ atom names occurring in the rules of P

    for each π in atoms($(\!|P|\!)$) **do**

        if $(\!|P|\!)(\pi) \notin F$ **then**

            $D := D \cup \{\pi\}$

            for each $i \in \{1, \ldots, \mathsf{num}(\pi)\}$ **do**

                $\pi' = \mathsf{atom}(\pi, i)$

                if $(\!|P|\!)(\pi') \in F$ or membrane(π) $\neq$ membrane(π') **then**

                    $A := A \cup \{(\mathsf{membrane}(\pi), \texttt{\#}, \{(1, (\!|P|\!)(\pi, i))\})\}$

                end if

            end for

        end if

    end for

    for each membrane π_M in $(\!|P|\!)$ **do**

        if atoms($(\!|P|\!), \pi_M) \subseteq D$ **then**

            $A := A \cup \{(\pi_M, \texttt{\#}, \emptyset)\}$

        end if

    end for

    return updated P by adding A and deleting D

end function

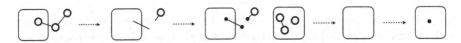

Fig. 9. UPE keeps a link crossing a membrane and terminates the end-points of free links with # (drawn as ●). Atoms except # are drawn as circles.

Fig. 10. UPE inserts a special nullary atom # (drawn as ●) into a membrane emptied during UPE.

does not delete any atoms whose names occur in the rewrite rules of P. That is, a subgraph of P that matches some rewrite rule of P remains unchanged.

What UPE is able to abstract away is not limited to atom names; UPE can also abstract graph structures as in Figs. 9 and 10.

5.2   UPE for State Spaces

We extend the abstraction function UPE to take state spaces as well. Given a state space $[\![P]\!] = (S_P, R_P, P)$, UPE($[\![P]\!]$) $= (S_P^{\#}, R_P^{\#}, \mathsf{UPE}(P))$ denotes a state space reduced by UPE, where $S_P^{\#} := \{\mathsf{UPE}(s) \mid s \in S_P\}$ and $R_P^{\#} :=$ $\{(\mathsf{UPE}(s), \mathsf{UPE}(t)) \mid (s, t) \in R_P\}$. The following theorem states that UPE preserves transitions in $[\![P]\!]$, i.e., UPE is a graph homomorphism for state spaces being considered.

Theorem 2. $\forall s, t \in S_P, (s, t) \in R_P \implies (\mathsf{UPE}(s), \mathsf{UPE}(t)) \in R_P^{\#}$.

Now we show an important property of UPE.

Theorem 3. $\mathsf{UPE}(\llbracket P \rrbracket) = \llbracket \mathsf{UPE}(P) \rrbracket$.

This theorem is very important in practice because applying UPE to all states is costly in constructing state space. Theorem 3 shows that applying UPE to the initial state once is sufficient to reduce state space.

The structural congruence relation can make the size of $\mathsf{UPE}(\llbracket P \rrbracket)$ defined above smaller than the size of $\llbracket P \rrbracket$. To show that, we first prove that UPE preserves the structural congruence relation.

Theorem 4. $\forall s, t \in S_p, \ s \equiv t \implies \mathsf{UPE}(s) \equiv \mathsf{UPE}(t)$

The next theorem is an almost immediate consequence.

Theorem 5. For $\llbracket P \rrbracket / {\equiv} = (S_{P/\equiv}, R_{P/\equiv}, P)$ and $\mathsf{UPE}(\llbracket P \rrbracket / {\equiv}) = (S^{\#}_{P/\equiv}, R^{\#}_{P/\equiv}, P^{\#})$, we have $|S^{\#}_{P/\equiv}| \leq |S_{P/\equiv}|$, i.e., UPE does not increase the number of states of $\llbracket P \rrbracket / {\equiv}$.

How much smaller $|S^{\#}_{P/\equiv}|$ becomes compared to $|S_{P/\equiv}|$ depends on individual models. We will study the effect of UPE in Sect. 6.

5.3   Soundness of UPE

A model is a state space including a labeling function L mentioned in Sect. 3.3. Since UPE can be regarded as an abstraction function, we discuss the soundness of UPE for LTL model checking in terms of abstract interpretation. Given a model $\llbracket P \rrbracket = (S_P, R_P, P, L)$ and its abstract model $\mathsf{UPE}(\llbracket P \rrbracket) = (S^{\#}_P, R^{\#}_P, \mathsf{UPE}(P), L')$, UPE is called sound iff the following two conditions hold [4]:

$$\forall s, t \in S_P, \ (s, t) \in R_P \implies (\mathsf{UPE}(s), \mathsf{UPE}(t)) \in R^{\#}_P$$
$$\forall s \in S_P, \ L(s) = L'(\mathsf{UPE}(s))$$

Note that $\llbracket P \rrbracket$ and $\mathsf{UPE}(\llbracket P \rrbracket)$ uses the same atomic propositions, which means that $L : S_P \to 2^{AP}$ and $L' : S^{\#}_P \to 2^{AP}$ has the same codomain.

UPE must preserve labeling functions to satisfy the conditions above. Accordingly, we modify UPE and add atom names occurring in atomic propositions in ϕ to the set F (in Algorithm 1) of atom names occurring in the original UPE. This prevents atoms which may affect the truth/falsity of atomic propositions from being abstracted away. That is, this modification enables UPE to equalize the results of labeling functions. It is clear that the modified UPE retains the properties of UPE shown as Theorems 2–5.

Theorem 6. Given a process P and an LTL formula ϕ, the modified UPE satisfies $\mathsf{UPE}(\llbracket P \rrbracket) \models \phi \implies \llbracket P \rrbracket \models \phi$.

6   Experiments

In order to confirm the effect of our proposed method, we measured the improvement of the number of states for some benchmark programs. We created LMNtal programs by translating popular concurrent algorithms in the textbook [2] and constructed their state spaces using SLIM. The translation was done manually but in a straightforward manner that could be automated without difficulty. Dekker, Peterson, Doran-Thomas, and Udding's starvation-free algorithms are algorithms for mutual exclusion of critical sections. The dining philosophers problem is a model with five philosophers and a fork placed between each pair of adjacent philosophers. When all the philosophers pick up their left forks first, a deadlock may occur. However, if one philosopher is perverse and picks up the right fork first, no deadlock will occur. We evaluated both the deadlock and non-deadlock versions straightforwardly translated from the textbook description of the algorithm. Note that they are less abstract, and have larger state space, than the program in Fig. 2 that fully exploits the features of LMNtal.

These algorithms have potential symmetry because individual processes in an original procedural algorithm are almost identical. However, processes symmetry does not immediately appear because variables and process names differ.

We experimented on Dekker, Peterson, and Doran-Thomas algorithms with two processes, Udding's algorithm with three processes, and the dining philosophers problem with five processes. The results are shown in Table 1. UPE reduced the number of states by extracting the essential behavior and the symmetry of a model. The exception was the non-deadlocking dining philosophers model for which changing the behavior of one philosopher collapsed the symmetry.

Table 1. The effect of UPE on the number of states

Problem	# of States	# of States (UPE)
Dekker	364	182
Peterson	190	95
Doran-Thomas	576	288
Udding's (3 processes)	7619	1478
Philosophers	16805	3365
Philosophers (no Deadlock)	16806	16806

The effect of UPE with different numbers of processes was measured for Udding's algorithm and the dining philosophers problem. The results are shown in Figs. 11 and 12. In Udding's algorithm, UPE reduced the number of states to nearly $1/N!$ times, N being the number of processes. This is because Udding's algorithm forms a star shape with semaphore variables in the center, and all permutations of the N processes become isomorphic. Dekker, Peterson, and Doran-Thomas algorithms also have a star topology, and the number of states of their models will be reduced to $1/N!$ by UPE. However, in the dining philosophers problem, UPE reduced the number of states only to $1/N$ of the non-UPE case because philosophers and forks form a ring topology rather than a star topology. Thus the effect of UPE depends on the symmetry of the topology of the model.

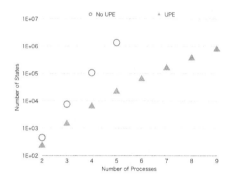
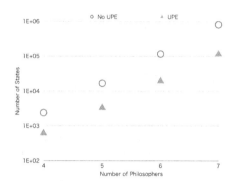

Fig. 11. Number of states of Udding's starvation-free algorithm

Fig. 12. Number of states of the dining philosophers problem

7   Conclusion and Future Work

In this paper, we focused on the symmetry of graph structures in model checking and proposed an abstraction method for strengthening the symmetry to reduce the size of state spaces. Our method makes models more symmetric based on equivalences built into the modeling language. We have shown that the method preserves soundness with respect to model checking. Our experiments showed that the proposed method reduces the size of state spaces when they are potentially symmetric. It is important to note that our rather simple UPE algorithm effectively introduced an abstraction method to an existing and working language and system. This is achieved by reducing an equivalence relation introduced by the abstraction to an equivalence relation inherent in the source language.

For future work, it is interesting to consider predicate abstraction in graph rewriting systems. Predicate abstraction is a successful technique [11] that divides domains of variables in programs depending on the truth values of predicates. Rewrite rules describe the behavior of models in graph rewriting systems and can also be considered as constraints on graphs. We used rather limited information of rewrite rules for our abstraction, but obtaining graph constraints by analyzing rewrite rules will enable more powerful abstraction.

Acknowledgments. The authors would like to thank anonymous reviewers for their useful comments. This work was partially supported by Grant-in-Aid for Scientific Research (B) JP18H03223, JSPS, Japan.

References

1. Backes, P., Reineke, J.: Analysis of infinite-state graph transformation systems by cluster abstraction. In: D'Souza, D., Lal, A., Larsen, K.G. (eds.) VMCAI 2015. LNCS, vol. 8931, pp. 135–152. Springer, Heidelberg (2015). https://doi.org/10.1007/978-3-662-46081-8_8
2. Ben-Ari, M.: Principles of Concurrent and Distributed Programming. Addison-Wesley, Boston (2006)
3. Clarke, E.M., Enders, R., Filkorn, T., Jha, S.: Exploiting symmetry in temporal logic model checking. Form. Methods Syst. Des. **9**(1), 77–104 (1996)
4. Clarke, E.M., Grumberg, O., Long, D.E.: Model checking and abstraction. ACM Trans. Program. Lang. Syst. **16**(5), 1512–1542 (1994)
5. Donaldson, A.F., Miller, A.: A computational group theoretic symmetry reduction package for the SPIN model checker. In: Johnson, M., Vene, V. (eds.) AMAST 2006. LNCS, vol. 4019, pp. 374–380. Springer, Heidelberg (2006). https://doi.org/10.1007/11784180_29
6. Emerson, E.A., Havlicek, J.W., Trefler, R.J.: Virtual symmetry reduction. In: Proceedings o LICS 2000, pp. 121–131. IEEE Computer Society (2000)
7. Emerson, E.A., Sistla, A.P.: Symmetry and model checking. Form. Methods Syst. Des. **9**(1–2), 105–131 (1996)
8. Feret, J.: An algebraic approach for inferring and using symmetries in rule-based models. Electron. Notes Theor. Comput. Sci. **316**, 45–65 (2015)
9. Ghamarian, A.H., de Mol, M., Rensink, A., Zambon, E., Zimakova, M.: Modelling and analysis using GROOVE. STTT **14**(1), 15–40 (2012)
10. Gocho, M., Hori, T., Ueda, K.: Evolution of the LMNtal runtime to a parallel model checker. Comput. Softw. **28**(4), 4_137–4_157 (2011)
11. Graf, S., Saidi, H.: Construction of abstract state graphs with PVS. In: Grumberg, O. (ed.) CAV 1997. LNCS, vol. 1254, pp. 72–83. Springer, Heidelberg (1997). https://doi.org/10.1007/3-540-63166-6_10
12. Holzmann, G.J.: The SPIN Model Checker: Primer and Reference Manual. Addison-Wesley Professional, Boston (2003)
13. Jensen, K.: Condensed state spaces for symmetrical coloured Petri Nets. Form. Methods Syst. Des. **9**(1–2), 7–40 (1996)
14. Junttila, T.: On the symmetry reduction method for Petri Nets and similar formalisms. Ph.D. thesis, Helsinki University of Technology (2003)
15. Miller, A., Donaldson, A.F., Calder, M.: Symmetry in temporal logic model checking. ACM Comput. Surv. **38**(3), 8 (2006)
16. Milner, R.: The Space and Motion of Communicating Agents. Cambridge University Press, Cambridge (2009)
17. Norris, I.P.C., Dill, D.L.: Better verification through symmetry. Form. Methods Syst. Des. **9**(1), 41–75 (1996)
18. Rensink, A.: Isomorphism checking in GROOVE. Electron. Commun. EASST **1** (2006). https://doi.org/10.14279/tuj.eceasst.1.77
19. Rensink, A., Distefano, D.: Abstract graph transformation. Electron. Notes Theor. Comput. Sci. **157**(1), 39–59 (2006)
20. Rozenberg, G.: Handbook of Graph Grammars and Computing by Graph Transformation. World Scientific, Singapore (1997)
21. Schmidt, K.: Integrating low level symmetries into reachability analysis. In: Graf, S., Schwartzbach, M. (eds.) TACAS 2000. LNCS, vol. 1785, pp. 315–330. Springer, Heidelberg (2000). https://doi.org/10.1007/3-540-46419-0_22

22. Sistla, A.P., Godefroid, P.: Symmetry and reduced symmetry in model checking. In: Berry, G., Comon, H., Finkel, A. (eds.) CAV 2001. LNCS, vol. 2102, pp. 91–103. Springer, Heidelberg (2001). https://doi.org/10.1007/3-540-44585-4_9
23. Sistla, A.P., Gyuris, V., Emerson, E.A.: SMC: a symmetry-based model checker for verification of safety and liveness properties. ACM Trans. Softw. Eng. Methodol. **9**(2), 133–166 (2000)
24. Ueda, K.: Encoding distributed process calculi into LMNtal. Electron. Notes Theor. Comput. Sci. **209**, 187–200 (2008)
25. Ueda, K.: Encoding the pure lambda calculus into hierarchical graph rewriting. In: Voronkov, A. (ed.) RTA 2008. LNCS, vol. 5117, pp. 392–408. Springer, Heidelberg (2008). https://doi.org/10.1007/978-3-540-70590-1_27
26. Ueda, K.: LMNtal as a hierarchical logic programming language. Theor. Comput. Sci. **410**(46), 4784–4800 (2009)

Double-Pushout Rewriting in Context
Rule Composition and Parallel Independence

Michael Löwe[(✉)]

FHDW Hannover, Freundallee 15, 30173 Hanover, Germany
michael.loewe@fhdw.de

Abstract. Recently, we introduced *double-pushout rewriting in context* (DPO-C) as a conservative extension of the classical double-pushout approach (DPO) at monic matches. DPO-C allows non-monic rules such that the split and merge of items can be specified together with deterministic context distribution and joining. First results showed that DPO-C is practically applicable, for example in the area of model refactoring, and that the theory of the DPO-approach is very likely to carry over to DPO-C. In this paper, we extend the DPO-C-theory. We investigate rule composition and characterise parallel independence.

1   Introduction

The classical double-pushout approach to graph and model rewriting (DPO) [4,5] uses spans of monic morphisms as rewrite rules in order to make sure that rewrites are deterministic and reversible in any adhesive category [9]. Therefore, DPO-rules can only specify deletion and addition of vertices and edges in any graph-like category. We claim that, from the practical point of view, it is also worthwhile to admit non-monic rules which allow the split and merge of items.

An intuitive example for such a rule in the application area of model refactoring is depicted in Fig. 1.[1] It specifies the extraction of an abstract type out of a given type denoted by 1/2 in the rule's left-hand side. The rule's left-hand transition from L to K specifies that *all* incoming associations and outgoing inheritance relations shall be connected to the more abstract type 1 and that *all* outgoing associations and incoming inheritance relations shall be adjacent to the more concrete type 2. This means, that the complete context of the refactored item 1/2 is *distributed* to the two split particles; nothing in the context is copied nor deleted. The rule's right-hand transition from K to R adds the necessary new inheritance relation from the concrete to the abstract type.

In order to handle *all* adjacent edges of the item 1/2 correctly, we must bind them to the representatives in the rule's left hand side L. For this purpose we will borrow a mechanism from AGREE-rewriting [1]. Given a match m for the type 1/2 in a model M, we map the *complete* model M back to the rule's left-hand side L: all other types besides $m(1/2)$ are mapped to the frame of

---

[1] For more examples, see [11].

© Springer Nature Switzerland AG 2019
E. Guerra and F. Orejas (Eds.): ICGT 2019, LNCS 11629, pp. 21–37, 2019.
https://doi.org/10.1007/978-3-030-23611-3_2

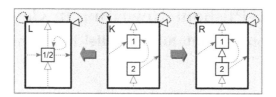

Fig. 1. Type extraction

L, adjacent associations and inheritance relations of $m(1/2)$ are mapped to the suitable representatives in L, and all other associations and inheritance relations are mapped to the loops on L's frame. Thus, the frames of L, K, and R in the rule stand for all other types that are not mentioned explicitly in the rule.

This inversion of the match offers a natural opportunity for negative application conditions like in [6]. In Fig. 1 for example, there is no inheritance loop on the type 1/2 such that the rule cannot be applied to ill-formed types with such a loop.[2] The negative application condition feature becomes more obvious, if we consider the rule in Fig. 2. It specifies the removal of a useless type. The type 1 is useless, since it has no outgoing and incoming associations and just one incoming inheritance relation from a type that has no other outgoing inheritance relations.

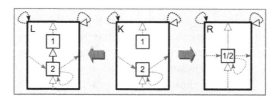

Fig. 2. Removal of useless abstraction

These example demonstrate, that splitting and merging items with controlled handling of all the context can be a powerful specification mechanism. Thus, it is worthwhile to extend DPO-rewriting by such mechanisms. We have introduced such an approach in [11,12] which we called *double pushout rewriting in context* DPO-C. In this paper, we provide more theoretical results for DPO-C which all turn out to be natural extensions of the corresponding DPO-results. The paper is organised as follows: Sect. 2 explains the categorical construction *partial arrow classifier* which allows the above mentioned inversion of the match. DPO-C is built on this feature. DPO-C is formally introduced in Sect. 3, which also contains a summary of the results obtained in [12]. Sections 4 and 5 present new theoretical results for rule composition/decomposition and parallel independence respectively. Finally, Sect. 6 discusses possible future research.

---

[2] Every well-structured object-oriented model shall be hierarchical.

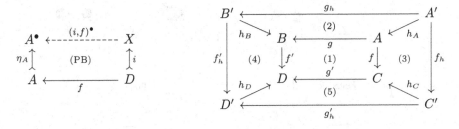

Fig. 3. Partial arrow classifier and commutative cube

2 Partial Arrow Classifier

For the above mentioned context matching, we need the categorical notion of *partial arrow classifiers*. In this section, we formally introduce them, recapitulate important and later used results, and illustrate by examples how classifiers are constructed and how they contribute to the required *inversion of matches*.

Definition 1 (Partial arrow classifiers). *A category has* partial arrow classifiers, *if for each object A there is monic morphism $\eta_A : A \rightarrowtail A^\bullet$ satisfying: For every pair $(i : D \rightarrowtail X, f : D \to A)$ of morphisms with monic i, there is a unique morphism $(i, f)^\bullet : X \to A^\bullet$ such that (i, f) is the pullback of $(\eta_A, (i, f)^\bullet)$, compare left part of Fig. 3. The morphism $(i, f)^\bullet$ is called the* totalisation *of (i, f). We abbreviate $(i, \mathrm{id}_A)^\bullet$ by $i^\bullet$.*

For categories with partial arrow classifiers, we know the following facts [8]:

Fact 2 (Properties of classified categories).

1. *For $d \circ a = c \circ b$ with monic a and c and arbitrary x such that $x \circ b$ is defined, $(a, x \circ b)^\bullet = (c, x)^\bullet \circ d$, if and only if (a, b) is pullback of (c, d). For the special case that x is the identity, we obtain: $(a, b)^\bullet = c^\bullet \circ d$, if and only if (a, b) is pullback of (c, d). The special case that b is the identity provides: $(a, x)^\bullet = (c, x)^\bullet \circ d$, if and only if (a, id) is pullback of (c, d).*
2. *All pushouts are hereditary: Pushout (f', g') of (g, f) in sub-diagram (1) in the right part of Fig. 3 is hereditary, if all commutative situations as in Fig. 3 where sub-diagrams (2) and (3) are pullbacks and h_B and h_C are monic satisfy: (f'_h, g'_h) is pushout of (g_h, f_h), if and only if sub-diagrams (4) and (5) are pullbacks and h_D is monic.*
3. *Pushouts preserve monomorphisms.*
4. *Pushouts of morphisms pairs (f, g) with monic g are pullbacks.*

In the rest of this section, we present some sample classifiers in the categories of sets, graphs, and simple object-oriented models.

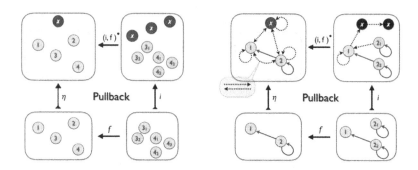

Fig. 4. Sample partial arrow classifiers in the categories Set and Graph

Example 3 (Classifier in the category Set of sets and maps). Given a set S, the classifier object is defined as $S^\bullet = S \cup \{\mathcal{X}\}$, i.e. the union of S with the final Set-object, and the classifier is the inclusion $\eta_S : S \rightarrowtail S \cup \{\mathcal{X}\}$. The left part of Fig. 4 shows a sample classifier for a four element set. It also shows an example how the unique morphism $(i, f)^\bullet : A \to S^\bullet)$ is constructed for a pair $(i : D \rightarrowtail A, f : D \to S)$: Every element with a pre-image under i is mapped as f does and all other elements are mapped to $\mathcal{X}$.

Example 4 (Classifier in the category G of graphs). For a given graph G, the classifier object $G^\bullet$ adds the final graph object, i.e. a vertex $\mathcal{X}$ with a loop, and an 'undefined' edge between every vertex in each direction. The classifier $\eta_G : G \rightarrowtail G^\bullet$ is again the resulting inclusion. The right part of Fig. 4 shows a sample classifier for a graph with two vertices, an edge between them, and a loop. The edges that the classifier adds are painted as dotted arrows.[3] Figure 4 also depicts a sample 'totalisation' for a pair (i, f): All vertices and edges with pre-image under i are mapped as f does, all other vertices are mapped to $\mathcal{X}$, and all other edges are mapped homomorphically to the uniquely determined suitable 'undefined' edge. Edges that are mapped to 'undefined' edges are painted as dotted arrows in Fig. 4.

Example 5 (Classifier in the category M of simple object-oriented models). Simple object-oriented models are algebras wrt. the following signature.[4]
```
M(odel) = sorts Type [painted as: □],
               Inheritance [painted as: ⇢],
               Association [painted as: →]
          opns  child, parent: Inheritance ⟶ Type
                owner, target: Association ⟶ Type
```
They are graph-like structures that distinguish two sorts of edges, namely associations and inheritance relations. Thus, the classifier for a model adds the final model, i.e. an 'undefined' type $\mathcal{X}$ together with an association and an inheritance loop, and an 'undefined' association and an 'undefined' inheritance relation between every type in each direction. Figure 5 depicts the classifier for a

---

[3] Two-headed arrows stand for a pair of arrows one in each direction.

[4] The examples in the introduction also use this underlying category.

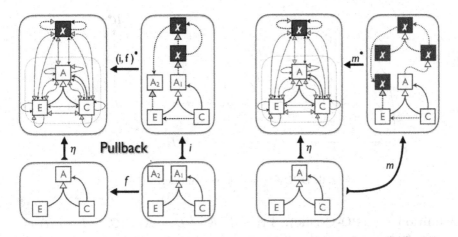

Fig. 5. Sample partial arrow classifier in the category M

model with three types, i.e. type A with two specialisations E and C, and an association from C to A. The left part of Fig. 5 shows a totalisation $(i, f)^\bullet$ for a pair (i, f) and the right part a totalisation $m^\bullet$ for a pair (m, id).[5]

The latter demonstrates the *inversion of matches* capability of totalisations. For a monic match $m : L \rightarrowtail M$, the totalisation $m^\bullet : M \to L^\bullet$ binds *all* items in M to suitable representatives in $L^\bullet$: All items in L are identically reflected, since $m^\bullet \circ m = \eta_L$, all associations and inheritance relations adjacent to items in L are bound to a unique structural compatible 'undefined' association or inheritance relation in $L^\bullet$, and all other context is mapped to the final model in $L^\bullet$.

3 DPO-Rewriting in Context

This section presents DPO-C by summarising definitions and results of [11].

Assumption 6 (Basic category). *For the rest of the paper, we assume an adhesive category with partial arrow classifiers.*

A category is *adhesive* if (i) it has all pullbacks and (ii) it has pushouts along monomorphisms which are all van-Kampen squares. A pushout (f', g') of (g, f) as (1) in the right part of Fig. 3 is van-Kampen, if, for every commutative diagram as depicted in Fig. 3 in which sub-diagrams (2) and (3) are pullbacks, (f_h', g_h') is pushout of (g_h, f_h), if and only if sub-diagrams (4) and (5) are pullbacks.

---

[5] Again, the associations and inheritance relations the classifier adds and the associations and inheritance relations which are mapped to them by the totalisation are painted as dotted arrows.

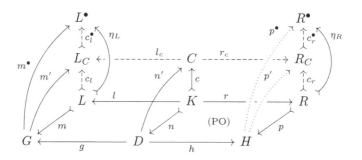

Fig. 6. DPO-C rule, match, and derivation

Definition 7 (DPO-C-rule). *A rule $(l : K \to L, c : K \rightarrowtail C, r : K \to R)$ is a triple of morphisms such that the context specification c is monic and, given the pushouts $(c_l : L \rightarrowtail L_C, l_c : C \to L_C)$ and $(c_r : R \rightarrowtail R_C, r_c : C \to R_C)$ of (l, c) and (r, c) respectively, the morphisms $c_l^\bullet$ and $c_r^\bullet$ are monic, compare Fig. 6.[6]*

The special rule format makes sure that items in L_C which are not in L, have a 'unique pre-image' under l_c. Thus, DPO-C does not allow any copying or deletion of contexts. Context can only be distributed to split particles. The symmetric restriction of the right side will ensure reversibility of rewrites.

Definition 8 (DPO-C-match and -derivation). *Given rule $\sigma = (l : K \to L, c : K \rightarrowtail C, r : K \to R)$, a monomorphism $m : L \rightarrowtail G$ is a match, if the following match condition is satisfied: The morphism $m^\bullet : G \to L^\bullet$ factors through L_C, i. e. there is $m' : G \to L_C$ such that $c_l^\bullet \circ m' = m^\bullet$.*

A derivation with rule σ at match m is constructed as follows, compare Fig. 6:

1. *Construct pullback $(g : D \to G, n' : D \to C)$ of $(m' : G \to L_C, l_c : C \to L_C)$.*
2. *Let $n : K \rightarrowtail D$ be the unique mediating morphism for this pullback for $(m \circ l, c)$. By pullback decomposition and Fact 2(4) for pushout (l_c, c_l), (l, n) is pullback of (g, m). Since pullbacks preserve monomorphisms, n is monic.*
3. *Construct pushout $(h : D \to H, p : R \rightarrowtail H)$ of $(n : K \rightarrowtail D, r : K \to R)$. The morphism p is monic by Fact 2(3).*

Remarks. Note that we restrict matches to monomorphisms. The morphism m' which satisfies the matching condition is unique, if it exists, since $c_l^\bullet$ is monic. The morphism n can be constructed in Step 2 of Definition 8, since $c_l^\bullet \circ m' \circ m = m^\bullet \circ m = \eta_L = c_l^\bullet \circ c_l$ implies $m' \circ m = c_l$ due to $c_l^\bullet$ being monic.

Fact 9 (Rewrite properties). *The sub-diagrams of a derivation with rule $\sigma = (l, c, r)$ at match m as depicted in Fig. 6 have the following properties:*

1. *(m, id_L) and (n, id_K) are pullbacks of (m', c_l) and (n', c) respectively.*
2. *(m, g) and (m', l_c) are pushouts of (l, n) and (g, n') respectively.*

---

[6] The pushout morphisms c_l and c_r are monic by Fact 2 (3).

3. *Because there is unique morphism $p' : H \to R_C$ for pushout (p, h) such that*
$p' \circ p = c_r$ *and* $p' \circ h = r_c \circ n'$,
(a) (h, n') and (p, id_R) are pullbacks of (r_c, p') and (p', c_r) respectively,
(b) $c_r^\bullet \circ p' = p^\bullet$, and
(c) (r_c, p') is pushout of (n', h).

DPO-C rewrites deterministic up to isomorphism justifying a special notation:

Notation 10 (Rewrite). *In a derivation with rule σ at match m as in Definition 8, the result H is denoted by $\sigma@m$, the span (g, h) is called the* trace, *written $\sigma \langle m \rangle$, and morphism p constitutes the* co-match, *written $m \langle \sigma \rangle$.*

Every DPO-C-rewrite is *reversible*: If we denote the inverse rule for a rule (l, c, r) by $\sigma^{-1} = (r, c, l)$ and the inverse trace for a trace (g, h) by $(g, h)^{-1} = (h, g)$, then, for any trace $\sigma \langle m \rangle$ and co-match $m \langle \sigma \rangle$, there is a trace $\sigma^{-1} \langle m \langle \sigma \rangle \rangle$ and co-match $m \langle \sigma \rangle \langle \sigma^{-1} \rangle$ such that $\sigma^{-1} \langle m \langle \sigma \rangle \rangle = \sigma \langle m \rangle^{-1}$ and $m \langle \sigma \rangle \langle \sigma^{-1} \rangle = m$.

DPO-C rewriting is a *conservative extension of the DPO approach* with linear rules at monic matches: if we define the DPO-C-simulation of a left- and right-linear DPO-rule $\varrho = (l : K \rightarrowtail L, r : K \rightarrowtail R)$ by $\sigma_\varrho = (l, \eta_K, r)$, then, for any match $m : L \rightarrowtail G$, $\varrho \langle m \rangle = \sigma_\varrho \langle m \rangle$ and $m \langle \varrho \rangle = m \langle \sigma_\varrho \rangle$.[7]

4   Rule Composition and Decomposition

The analysis with respect to rule composition and decomposition investigates how more complex rules can be built up from and decomposed into more elementary parts. In the DPO-approach two rules $\varrho_1 = (l_1 : K_1 \rightarrowtail L_1, r_1 : K_1 \rightarrowtail R_1)$ and $\varrho_2 = (l_2 : K_2 \rightarrowtail L_2, r_2 : K_2 \rightarrowtail R_2)$ can be composed, if $R_1 = L_2$. The composition is given by the standard notion of span composition[8] by constructing the pullback of the first rule's right-hand side and the second rule's left-hand side morphism, i.e. $\varrho_2 \circ \varrho_1 = (l_1 \circ l_2' : K_{12} \to L_1, r_2 \circ r_1' : K_{12} \to R_2)$ where (l_2', r_1') is pullback of (r_1, l_2), compare upper part in Fig. 7. The composition $\varrho_2 \circ \varrho_1$ is a span of monomorphisms, since pullbacks preserve monomorphisms.

In the DPO-approach, it is easy to show that rule composition leads to trace composition in derivations, i.e. $(\varrho_2 \circ \varrho_1) \langle m \rangle = \varrho_2 \langle m \langle \varrho_1 \rangle \rangle \circ \varrho_1 \langle m \rangle$ and $m \langle \varrho_2 \circ \varrho_1 \rangle = m \langle \varrho_1 \rangle \langle \varrho_2 \rangle$ for any match m. Consider Fig. 7 where $\varrho_1 = (l_1, r_1)$, $\varrho_2 = (l_2, r_2)$, $\varrho_2 \circ \varrho_1 = (l_1 \circ l_2', r_2 \circ r_1')$, $\varrho_1 \langle m_1 \rangle = (g_1, h_1)$, $m_1 \langle \varrho_1 \rangle = m_2$, and $\varrho_2 \langle m_1 \langle \varrho_1 \rangle \rangle = (g_2, h_2)$. Construct (g_2', h_1') as pullback of (h_1, g_2) which provides morphism p_{12} making the diagram commutative. By Fact 2(4), (r_1, n_1) and (l_2, n_2) are pullbacks. By pullback composition and decomposition, (r_1', p_{12}) and (l_2', p_{12}) are pullbacks as well. Hereditariness Fact 2(2) guarantees that (n_1, g_2') and (n_2, h_1') are pushouts. Finally, pushout composition produces the two required pushouts for the application of the composed rule, namely $(m_1, g_1 \circ g_2')$ and $(p_2, h_2 \circ h_1')$.

---

[7] If (m, g) and (p, h) are pushouts of (n, l) resp. (n, r) in a DPO-derivation with rule $\varrho = (l, r)$ at match m, we denote the trace (g, h) by $\varrho \langle m \rangle$ and co-match p by $m \langle \varrho \rangle$.

[8] See [13] for composition of partial maps and especially composition of monic spans.

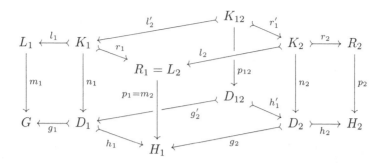

Fig. 7. Rule composition in DPO

This analysis shows that most rewrite rules are compositions of more elementary rules. In the category of graphs for example, every DPO-rule can be built up by elementary rules which add or delete a single item, i. e. either an edge or a vertex without adjacent edges.[9]

For the composition of DPO-C-rules, coincidence of the first rule's right-hand side with the second rule's left-hand side is not sufficient. We have to take the context specification into account as well.

Definition 11 (Rule composition). *Two DPO-C rules $\sigma_1 = (l_1, c_1, r_1)$ and $\sigma_2 = (l_2, c_2, r_2)$ can be composed, if $c_r^1 = c_l^2$. The composition $\sigma_2 \bullet \sigma_1 = (l_1 \circ l_2', c_{12}, r_2 \circ r_1')$ is defined by the pullback (l_2', r_1') of (r_1, l_2), the pullback (l_c^{12}, r_c^{12}) of (r_c^1, l_c^2), and the morphism c_{12} making the diagram commutative, compare upper part of Fig. 8.*

Proposition 12 (Composed rule). *Rule composition is well-defined.*

Proof. We must show that the composed rule satisfies the requirements of Definition 7. Consider the upper part of Fig. 8. By Fact 2(4), (c_1, r_1) and (c_2, l_2) are pullbacks. By pullback composition and decomposition, (l_2', c_{12}) and (r_1', c_{12}) are pullbacks and, since pullbacks preserve monomorphisms, c_{12} is monic. Now, the van-Kampen-property guarantees that (l_c^{12}, c_1) and (r_c^{12}, c_2) are pushouts of (c_{12}, l_2') and (c_{12}, r_1') respectively. Finally, pushout composition provides the two pushouts $(c_l^1, l_c^1 \circ l_c^{12})$ and $(c_r^2, r_c^2 \circ r_c^{12})$ of $(l_1 \circ l_2', c_{12})$ and $(r_2 \circ r_1', c_{12})$ respectively which satisfy the requirements of Definition 7.   □

As in the DPO-approach, composition of rules is consistent with derivations.

Proposition 13 (Composition of derivation). *Given a composition $\sigma_2 \bullet \sigma_1$ of DPO-C rules σ_1 and σ_2 and a derivation $\sigma_1 @ m_1$ with co-match $m_1 \langle \sigma_1 \rangle$, there is a derivation $(\sigma_2 \bullet \sigma_1) @ m_1$, if and only if there is a derivation $\sigma_2 @ m_1 \langle \sigma_1 \rangle$. In this case, traces and co-matches of the two derivations satisfy $(\sigma_2 \bullet \sigma_1) \langle m_1 \rangle = \sigma_2 \langle m_1 \langle \sigma_1 \rangle \rangle \circ \sigma_1 \langle m_1 \rangle$ and $m_1 \langle \sigma_2 \bullet \sigma_1 \rangle = m_1 \langle \sigma_1 \rangle \langle \sigma_2 \rangle$.*

---

[9] To my knowledge, rule and trace composition has never been investigated in isolation in the DPO-approach. Some aspects are handled within the concurrency theorem [4].

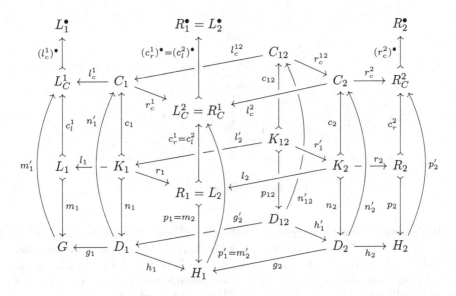

Fig. 8. Rule composition in DPO-C

Proof. Consider Fig. 8 which depicts the composition $(l_1 \circ l_2', c_{12}, r_2 \circ r_1')$ of the rules $\sigma_1 = (l_1, c_1, r_1)$ and $\sigma_2 = (l_2, c_2, r_2)$ in the upper and the derivation with σ_1 at match m_1 by the pullback (g_1, n_1') and pushout (p_1, h_1) in the left part.

"$\Leftarrow$": If we have the derivation with rule σ_2 at match $p_1 = m_2$, there is pullback (g_2, n_2') and pushout (p_2, h_2). Construct pullback (g_2', h_1') of (h_1, g_2) which provides morphisms p_{12} and n_{12}' making the resulting diagram commutative.

Pullbacks (g_2', h_1') and (g_2, n_2') compose, such that (g_2', n_{12}') is pullback by pullback decomposition, since (l_c^{12}, r_c^{12}) is pullback as well. By pullback composition, $(g_1 \circ g_2', n_{12}')$ is the left-hand side of the derivation with $\sigma_2 \bullet \sigma_1$ at match m_1.

Since (l_2, n_2) is pullback by Definition 8, pullback composition and decomposition guarantee that (l_2', p_{12}) is pullback. By van-Kampen, (h_1', n_2) is pushout of (p_{12}, r_1'). Pushout composition provides $(p_2, h_2 \circ h_1')$ as pushout of $(r_2 \circ r_1', p_{12})$ which is the right-hand side of the derivation with $\sigma_2 \bullet \sigma_1$ at match m_1.

"$\Rightarrow$": If we have the derivation with rule $\sigma_2 \bullet \sigma_1$ at match m_1, there is pullback $(g_1 \circ g_2', n_{12}')$ and pushout $(p_2, h_2 \circ h_1')$ which can both be decomposed into pullbacks (g_1, n_1') and (g_2', n_{12}') resp. into pushouts (n_2, h_1') and (p_2, h_2) the second of which is the right-hand side of the derivation with σ_2 at match m_2. The pushout (n_2, h_1') provides two mediating morphisms g_2 and n_2' and the pullback (g_2', n_{12}') provides the mediating morphism p_{12} making the diagram commutative. Now, (l_2', r_1') and (l_2', p_{12}) are pullbacks[10] and (m_2, h_1) and (n_2, h_1') are pushouts such that the van-Kampen-property provides pullbacks (g_2', h_1') and (l_2, n_2) the first of which is the composition of the traces (g_1, h_1) for σ_1 and (g_2, h_2) for σ_2.

---

[10] The latter by pullback decomposition of (c_{12}, l_2').

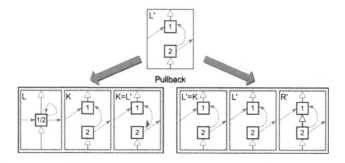

Fig. 9. Rule decomposition

It remains to show that (g_2, n_2') is pullback of (l_c^2, p_1'). The pullbacks (g_2', n_{12}'), (id_{K_1}, n_1), $(\mathrm{id}_{K_{12}}, p_{12})$, and the trivial pullback $(\mathrm{id}_{k_{12}}, l_2')$ of $(\mathrm{id}_{K_1}, l_2')$ together with the pushout (c_1, l_c^{12}) provide pushout (n_1, g_2') of (l_2', p_{12}) by van-Kampen. By pushout composition/decomposition, we conclude that (m_2, g_2) is pushout of (l_2, n_2). Pullbacks $(\mathrm{id}_{K_{12}}, p_{12})$ of $(c_{12}, n_{12}')^{11}$ and $(\mathrm{id}_{K_{12}}, r_1')$ of $(r_1', \mathrm{id}_{K_2})$ together with pushouts (c_2, r_c^{12}) and (n_2, h_1') lead to pullbacks (n_{12}', h_1') of (r_c^{12}, n_2') and (n_2, id_{k_2}) of (c_2, n_2') again by van-Kampen. A final application of the van-Kampen-property to pullback (n_2, id_{k_2}), trivial pullback (id_{K_2}, l_2) of (id_{L_2}, l_2) and pushouts (l_c^2, c_l^2) and (m_2, g_2) provides pullback (g_2, n_2') of (m_2', l_c^2). $\qquad\square$

Corollary 14 (Decomposition of derivations). *Derivations with a composed rule $\sigma \bullet \sigma'$ can be decomposed into derivations with the component rules, i.e. $(\sigma \bullet \sigma') \langle m \rangle = \sigma \langle m \langle \sigma' \rangle \rangle \circ \sigma' \langle m \rangle$ for every match m for $\sigma \bullet \sigma'$.*

Proof. Consequence of Property 13 and the fact that every $\sigma \bullet \sigma'$-match is σ'-match. $\qquad\square$

This analysis demonstrates that DPO-C derivations possess the same composition and decomposition properties as simple pushouts in arbitrary categories. Furthermore, Property 13 and Corollary 14 show that more complex DPO-C rules can be built up by more elementary rules. In the category of graphs, these elementary rules consist of the rules that add or delete a single item, which are well-known from DPO-rewriting, rules that split a single edge or vertex into two items (with context distribution in the case of vertices), and rules which merge two items.

Example 15 (Decomposition). The rule "type extraction" in Fig. 1 in the introduction formulates two actions, namely the *split* of class $1/2$ into classes 1 and 2 with a distribution of all adjacent context edges and the *addition* of a single edge between classes 1 and 2. We can make these elementary actions explicit by decomposing the rule into two simpler rules. These rules are depicted in Fig. 9.

---

[11] See Rewrite Property 9(1).

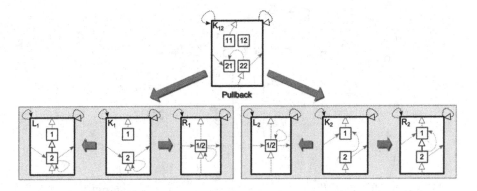

Fig. 10. Rule composition: reusing useless abstraction

In the left lower part is the rule for the split. Its right-hand side is the identity. In the right lower part is the rule for the addition. Its left-hand side is the identity. The original rule is re-obtained by the pullback (pair of identity morphisms here) of the first rule's right-hand side and the second rule's left-hand side.

Decomposing rule "removal of useless abstraction" in Fig. 2 will also result in two elementary actions, namely a *deletion* of a single edge and a *merge* of 2 vertices.

Example 16 (Composition). The two rules "removal of useless abstraction" and "type extraction" in the introduction, compare Figs. 2 and 1 respectively, can be composed, since the right-hand side of the first rule coincides with the left-hand side of the second rule. Figure 10 depicts both rules and their composition. The result is a rule "reusing useless abstraction" whose overall effect is the redistribution of the context associations of class 2 to classes 1 (in-coming) and 2 (out-going). The actions of this rule are: (1) delete the inheritance relation, (2a) split class 1 into classes 11 and 12 and connect the only possible sort of context edges to class 11, (2b) split class 2 into classes 21 and 22 and distribute the context edges such that all in-coming ones get connected to 21 and all out-going ones are connected to 22, (3) merge the classes 11 and 21 to 1 as well as 12 and 22 to 2, and (4) add an inheritance relation between sub-class 2 and super-class 1.

5   Analysis and Characterisation of Parallel Independence

Parallel independence analysis investigates the conditions under which two rewrites of the same object can be performed in either order and produce the same result. Essential for the theory is the notion of *residual match*: Under which conditions are two matches $m_G : L \rightarrowtail G$ and $m_H : L \rightarrowtail H$ for a rule's left-hand side L *the same match*, if there is a trace $(g : D \to G, h : D \to H)$? The DPO-answer is: m_G and m_H are the same, if there is $m_D : L \rightarrowtail D$ with $g \circ m_D = m_G$ and $h \circ m_D = m_H$, compare [7]. This answer is not sufficient for DPO-C, since

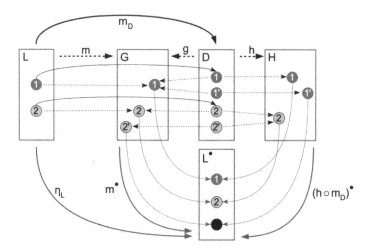

Fig. 11. Fake residual

we need to take the context matches into account as well, i. e. $m_G^\bullet$, $m_D^\bullet$, and $m_H^\bullet$ shall classify the 'same objects' in the same way. An example for inconsistent classification is given by Example 17.

Example 17 (Fake residual). Consider Fig. 11 which, in the category Set, shows a trace $(g : D \to G, h : D \to H)$, a match $m : L \rightarrowtail G$, and a candidate map $m_D : L \rightarrowtail D$ for a residual of m. We have $g \circ m_D = m$, but the classification $m^\bullet$, $m_D^\bullet$, and $(h \circ m_D)^\bullet$ are inconsistent. The map $m_D^\bullet$ classifies the items 1' and 2' as undefined, while $g(1')$ and $h(2')$ are not classified as undefined by $m^\bullet$ and $(h \circ m_D)^\bullet$ resp. Thus, we neither have $m^\bullet \circ g = m_D^\bullet$ nor $(h \circ m_D)^\bullet \circ h = m_D^\bullet$.

This analysis demonstrates that $m_G^\bullet$, $m_D^\bullet$, and $m_H^\bullet$ classify the 'same objects' in the same way only if $m_G^\bullet \circ g = m_D^\bullet$ and $m_H^\bullet \circ h = m_D^\bullet$. By Fact 2 (1), this is equivalent to requiring that (id_L, m_D) is pullback of (m_G, g) and (m_H, h).[12]

Definition 18 (Residual). *Let* $(g : D \to G, h : D \to H)$ *be a trace of a direct derivation and* m *match for rule* σ *in* G. *A match* $m^{(g,h)}$ *for* σ *in* H *is the* residual *of* m *for trace* (g, h), *if there is morphism* m_D *from the left-hand side* L *of* σ *to* D *such that* (id_L, m_D) *is pullback of* (m, g) *and* $(m^{(g,h)}, h)$.[13]

The pullback properties uniquely determine *the* residual, if it exists. Two derivations are parallel independent, if there are mutual residuals.

Definition 19 (Parallel independence). *Two derivations with rules* σ_1 *and* σ_2 *at matches* m_1 *and* m_2 *resp. rewriting the same object are* parallel independent, *if* m_1 *has a residual for* $\sigma_2 \langle m_2 \rangle$ *and* m_2 *has a residual for* $\sigma_1 \langle m_1 \rangle$.

---

[12] This condition is identical to the one in [3].

[13] This notion of residual is a conservative extension of the notion for DPO-rewriting with monic rules at monic matches, since traces become monic in this case as well and it is easy to see that a triangle $g \circ m_D = m$ with monic g is a pullback diagram.

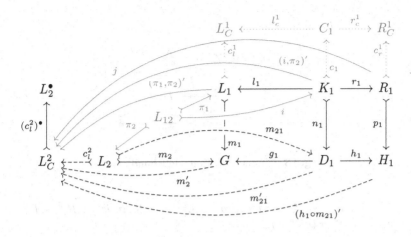

Fig. 12. Residual by independence

In [11,12], we proved that parallel independent derivations are confluent.

Theorem 20 (Confluence). *If derivations with rules σ_1 and σ_2 at matches m_1 and m_2 are parallel independent, then the derivations with the mutual residuals $m_1^{\sigma_2\langle m_2\rangle}$ and $m_2^{\sigma_1\langle m_1\rangle}$ produce the same result up to isomorphism, i. e. we can choose $\sigma_1@m_1^{\sigma_2\langle m_2\rangle} = \sigma_2@m_2^{\sigma_1\langle m_1\rangle}$. Additionally, the two traces of the two involved derivation sequences can be chosen, such that:*

$$\sigma_2\left\langle m_2^{\sigma_1\langle m_1\rangle}\right\rangle \circ \sigma_1\langle m_1\rangle = \sigma_1\left\langle m_1^{\sigma_2\langle m_2\rangle}\right\rangle \circ \sigma_2\langle m_2\rangle .$$

The parallel independence criterion of Definition 19 for confluence is not easy to check, since it requires the calculation of two complete derivations. Here, we seek for a better criterion that can be checked on the basis of the participating rules and matches only. Since the DPO-C approach to rewriting is much simpler than the general AGREE-framework, we do not need the heavy machinery of [10] for this purpose. For our analysis, the criteria in [2] provide a solid basis. We reduce the complexity for the confluence check by providing an easy to check characterisation for the existence of residuals.

Definition 21 (Independent match). *Given rule $\sigma_1 = (l_1 : K_1 \to L_1, c_1, r_1)$, the left-hand side $(c_l^2 : L_2 \rightarrowtail L_C^2, (c_l^2)^\bullet : L_C^2 \rightarrowtail L_2^\bullet)$ of another rule σ_2, two matches $m_1 : L_1 \rightarrowtail G$ and $m_2 : L_2 \rightarrowtail G$, and the pullback $(\pi_1 : L_{12} \rightarrowtail L_1, \pi_2 : L_{12} \rightarrowtail L_2)$ of the matches m_1 and m_2, then m_2 is independent of m_1, if there is morphism $i : L_{12} \rightarrowtail K_1$ satisfying:[14]*

1. *$r_1 \circ i$ is monic,*
2. *$(i, \mathrm{id}_{L_{12}})$ is pullback of (π_1, l_1) and $(r_1, r_1 \circ i)$, and*

---

[14] Compare Fig. 12.

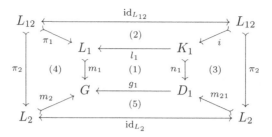

Fig. 13. Characterisation—left-hand side

3. $(r_1 \circ i, \pi_2)^\bullet : R_1 \to L_2^\bullet$ factors through L_C^2, i. e. there is $j : R_1 \to L_2^C$ such that $(r_1 \circ i, \pi_2)^\bullet = (c_l^2)^\bullet \circ j$.

Proposition 22 (Existence of residuals). A match m_2 for rule σ_2 is independent of match m_1 for σ_1, if and only if m_2 has a residual for trace $\sigma_1 \langle m_1 \rangle$.

Proof. "⇐": Consider Fig. 12. If m_2 has a residual for $\sigma_1 \langle m_1 \rangle$, there is m_{21} such that (i) $h_1 \circ m_{21}$ is monic, (ii) $(m_{21}, \text{id}_{L_2})$ is pullback of (m_2, g_1) and $(h_1 \circ m_{21}, h_1)$, and (iii) there is $(h_1 \circ m_{21})'$ such that $(c_l^2)^\bullet \circ (h_1 \circ m_{21})' = (h_1 \circ m_{21})^\bullet$.

Now, consider Fig. 13 where sub-diagrams (1), (4), and (5) are given as pullbacks. Construct pullback $(L_{12}, \pi_2, \text{id}_{L_{12}})$ of (π_2, id_{L_2}). This provides morphism $i : L_{12} \rightarrowtail K_1$ making the diagram commute. Since (4) and $(\pi_2, \text{id}_{L_{12}})$ as well as (1) are pullbacks, pullback composition and decomposition properties guarantee that (3) is pullback. Since (5)+(3) and (4) are pullbacks

$$(i, \text{id}_{L_{12}}) \text{ is pullback of } (\pi_1, l_1) \tag{1}$$

Now, consider Fig. 14 where sub-diagrams (1), (4), and (5) are given as pullbacks. Construct pullback (3) as (X, x, y) of $(p_1, h_1 \circ m_{21})$ such that $(h_1 \circ m_{21}) \circ x = p_1 \circ y$. This provides unique morphism $z : L_{12} \to X$ such that (2) becomes pullback as well, i. e. (i, z) is pullback of (y, r_1). Since (1)+(2) and (5) are pullbacks, the outer square in Fig. 14 is pullback, such that we can choose $X = L_{12}$, $z = \text{id}_{L_{12}}$, $y = r_1 \circ i$, and $x = \pi_2$. Thereby and by the fact that (3) is pullback, we obtain:

$$r_1 \circ i \text{ is monic and } (i, \text{id}_{L_{12}}) \text{ is pullback of } (r_1 \circ i, r_1) \tag{2}$$

Since $(r \circ i, \pi_2)$ is pullback of $(p_1, h \circ m_{21})$, Fact 2(1) provides $(r_1 \circ i, \pi_2)^\bullet = (h_1 \circ m_{21})^\bullet \circ p_1 \overset{\text{(iv)}}{=} (c_l^2)^\bullet \circ (h_1 \circ m_{21})' \circ p_1$. Thus, $(r \circ i, \pi_2)^\bullet$ factors through L_C^2 and there is j with

$$j = (h_1 \circ m_{21})' \circ p_1 \text{ and } (r \circ i, \pi_2)^\bullet = (c_l^2)^\bullet \circ j \tag{3}$$

Properties (1), (2), and (3) show that match m_2 is independent of match m_1.
"⇒": Consider again Fig. 12. If match m_2 is independent of match m_1, there are morphisms i and j, such that (iv) $r_1 \circ i$ is monic, (v) $(i, \text{id}_{L_{12}})$ is pullback of

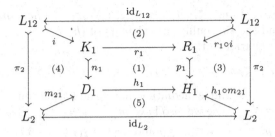

Fig. 14. Characterisation—right-hand side

(π_1, l_1) and $(r_1, r_1 \circ i)$, and (vi) $(r_1 \circ i, \pi_2)^\bullet = (c_l^2)^\bullet \circ j$. This implies $(i, \pi_2)^\bullet = (r_1 \circ i, \pi_2)^\bullet \circ r_1 = (c_l^2)^\bullet \circ (j \circ r_1)$ such that $(i, \pi_2)^\bullet$ factors through L_C^2. In Fig. 12, $(j \circ r_1)$ is denoted by $(i, \pi_2)'$.

Here, we have pullbacks (1), (2), and (4) and pushout (1) given in Fig. 13. Construct pullback (5) as (X, x, m_{21}) of (m_2, g_1). This provides unique morphism $y : L_{12} \to X$ making the diagram commutative and pullback (3) as (i, y) of (n_1, m_{21}). Since sub-diagram (1) is also pushout, the van-Kampen-property guarantees that (π_2, x) is pushout of $(\mathrm{id}_{L_{12}}, y)$. Since pushouts can be chosen to preserve identities[15], $x = \mathrm{id}_{L_2}$ and $y = \pi_2$. Thus,

$$(i, \pi_2) \text{ is pullback of } (n_1, m_{21}) \text{ and } (m_{21}, \mathrm{id}_{L_{12}}) \text{ is pullback of } (m_2, g_1) \quad (4)$$

In Fig. 14, sub-diagrams (1), (2), and (4) are given as pullbacks and (1) as pushout. Construct the pushout $(\mathrm{id}_{L_2}, \pi_2)$ of $(\pi_2, \mathrm{id}_{L_{12}})$. This provides unique morphism $x : L_2 \to H_1$ making the diagram commutative which means that $x = h_1 \circ m_{21}$. By hereditariness (Fact 2(2)), sub-diagram (3) is pullback and

$$h_1 \circ m_{21} \text{ is monic and } (m_{21}, \mathrm{id}_{L_2}) \text{ is pullback of } (h_1 \circ m_{21}, h_1). \quad (5)$$

With Eqs. (4) and (5), it remains to show that $(h_1 \circ m_{21})^\bullet$ factors through L_C^2. Since m_2 is match, we know that $m_2^\bullet$ factors through L_C^2, i.e. there is m_2' such that $m_2^\bullet = (c_l^2)^\bullet \circ m_2'$. By Fact 2(1) and $(m_{21}, \mathrm{id}_{L_{12}})$ being pullback of (m_2, g_1), $m_{21}^\bullet = m_2^\bullet \circ g_1 = (c_l^2)^\bullet \circ (m_2' \circ g_1)$ such that $m_{21}^\bullet$ factors through L_C^2, i.e. there is m_{21}' with $m_{21}^\bullet = (c_l^2)^\bullet \circ m_{21}'$ and $m_2' \circ g_1 = m_{21}'$. By Fact 2(1) and (i, π_2) being pullback of (n_1, m_{21}),

$$(c_l^2)^\bullet \circ m_{21}' \circ n_1 = m_{21}^\bullet \circ n_1 = (i, \pi_2)^\bullet = (c_l^2)^\bullet \circ (i, \pi_2)' \implies m_{21}' \circ n_1 = (i, \pi_2)'$$

Again by Fact 2(1) and (id, i) being pullback of $(r_1, r_1 \circ i)$

$$(c_l^2)^\bullet \circ (i, \pi_2)' = (i, \pi_2)^\bullet = (r_1 \circ i, \pi_2)^\bullet \circ r_1 = (c_l^2)^\bullet \circ j \circ r_1 \implies (i, \pi_2)' = j \circ r_1$$

---

[15] More precisely, pushouts preserve isomorphisms.

By these results, we obtain: $m_{21}^\bullet \circ n_1 = (r_1 \circ i, \pi_2)^\bullet \circ r_1$ and $m_{21}' \circ n_1 = j \circ r_1$. Two times Fact 2(1) and the two pullbacks (id, m_{21}) of $(h_1, h_1 \circ m_{21})$ and $(\pi_2, r_1 \circ i)$ of $(p_1, h_1 \circ m_{21})$[16] guarantee that

$$(h_1 \circ m_{21})^\bullet \circ p_1 = (r_1 \circ i, \pi_2)^\bullet \text{ and } (h_1 \circ m_{21})^\bullet \circ h_1 = m_{21}^\bullet$$

Thus, $(h_1 \circ m_{21})^\bullet$ is the unique mediating morphism for pushout (h_1, p_1) and the pair $(m_{21}^\bullet, (r_1 \circ i, \pi_2)^\bullet)$. Since we also have $m_{21}' \circ n_1 = j \circ r_1$, there is $(h_1 \circ m_{21})'$ with $(h_1 \circ m_{21})' \circ p_1 = (r_1 \circ i, \pi_2)' = j$ and $(h_1 \circ m_{21})' \circ h_1 = m_{21}'$. This provides

$$\left[(c_l^2)^\bullet \circ (h_1 \circ m_{21})' \right] \circ p_1 = (c_l^2)^\bullet \circ j = (r_1 \circ i, \pi_2)^\bullet \text{ and}$$

$$\left[(c_l^2)^\bullet \circ (h_1 \circ m_{21})' \right] \circ h_1 = (c_l^2)^\bullet \circ m_{21}' = m_{21}^\bullet.$$

Therefore, $(c_l^2)^\bullet \circ (h_1 \circ m_{21})'$ is also a mediating morphism for pushout (h_1, p_1) and the pair $(m_{21}^\bullet, (r_1 \circ i, \pi_2)^\bullet)$. Thus, $(c_l^2)^\bullet \circ (h_1 \circ m_{21})' = (h_1 \circ m_{21})^\bullet$ □

The notion of independent match is asymmetric: match m_1 for rule σ_1 can be independent of match m_2 for rule σ_2 while m_2 is dependent of m_1. This means that the derivation and trace $\sigma_2 \langle m_2 \rangle$ can be computed without destroying the match m_1, since there is the residual $m_1^{\sigma_2 \langle m_2 \rangle}$ and the trace $\sigma_1 \left\langle m_1^{\sigma_2 \langle m_2 \rangle} \right\rangle$ which produces the 'same' effect of σ_1 at match m_1 after σ_2 has been applied.

Theorem 23 (Characterisation of parallel independence). *Two derivations with rule σ_1 at match m_1 and rule σ_2 at match m_2 starting at the same object are parallel independent, if and only if m_1 and m_2 are mutual independent.*

Proof. Direct consequence of Proposition 22.

6   Conclusion

In this paper, we have shown two new results for DPO-C-rewriting. Both results are straightforward generalisations of corresponding results for the classical double-pushout approach (DPO) [4,5]. Thus, DPO-C proves again that it preserves all good properties of DPO, for example reversible and decomposable rules and derivations as well as derivation-independent characterisation of parallel independence, while enhancing the expressive power by non-monic rules. These rules allow item splitting and merging together with deterministic context-distribution and -join respectively. These features improve the applicability in some application areas, for example model refactorisation. DPO-C, however, does not allow "wild" actions, like copying and deleting complete graphs, as it is possible in AGREE-rewriting [1]. Thus, it is a "tamed AGREE-tiger".

From the theoretical point of view, there is a natural program for future research, namely the generalisation of other DPO-results to DPO-C for example with respect to sequential independence, concurrency, critical pair analysis, parallelism, and amalgamation. Besides that, future research can address the following issues:

---

[16] Pushout (p_1, h_1) of (r_1, n_1) is also pullback by Fact 2(4) and pullbacks compose.

– Comparison of the DPO-C-built-in negative application conditions to the well-known negative application conditions from the literature, e.g. [6],
– Comparison of DPO-C to other reversible approaches e.g. [3],
– Development of a clear and handy visual notation for the rules especially for the context specification, and
– Elaboration of bigger case studies e.g. in the field of model transformation.

References

1. Corradini, A., Duval, D., Echahed, R., Prost, F., Ribeiro, L.: AGREE – algebraic graph rewriting with controlled embedding. In: Parisi-Presicce, F., Westfechtel, B. (eds.) ICGT 2015. LNCS, vol. 9151, pp. 35–51. Springer, Cham (2015). https://doi.org/10.1007/978-3-319-21145-9_3
2. Corradini, A., et al.: On the essence of parallel independence for the double-pushout and sesqui-pushout approaches. In: Heckel, R., Taentzer, G. (eds.) Graph Transformation, Specifications, and Nets. LNCS, vol. 10800, pp. 1–18. Springer, Cham (2018). https://doi.org/10.1007/978-3-319-75396-6_1
3. Danos, V., Heindel, T., Honorato-Zimmer, R., Stucki, S.: Reversible sesqui-pushout rewriting. In: Giese, H., König, B. (eds.) ICGT 2014. LNCS, vol. 8571, pp. 161–176. Springer, Cham (2014). https://doi.org/10.1007/978-3-319-09108-2_11
4. Ehrig, H., Ehrig, K., Prange, U., Taentzer, G.: Fundamentals of Algebraic Graph Transformation. Springer, Heidelberg (2006). https://doi.org/10.1007/3-540-31188-2
5. Ehrig, H., Ermel, C., Golas, U., Hermann, F.: Graph and Model Transformation - General Framework and Applications. Monographs in Theoretical Computer Science. An EATCS Series. Springer, Heidelberg (2015). https://doi.org/10.1007/978-3-662-47980-3
6. Habel, A., Heckel, R., Taentzer, G.: Graph grammars with negative application conditions. Fundam. Inform. **26**(3/4), 287–313 (1996)
7. Habel, A., Müller, J., Plump, D.: Double-pushout graph transformation revisited. Math. Struct. Comput. Sci. **11**(5), 637–688 (2001)
8. Heindel, T.: Hereditary pushouts reconsidered. In: Ehrig, H., Rensink, A., Rozenberg, G., Schürr, A. (eds.) ICGT 2010. LNCS, vol. 6372, pp. 250–265. Springer, Heidelberg (2010). https://doi.org/10.1007/978-3-642-15928-2_17
9. Lack, S., Sobocinski, P.: Adhesive and quasiadhesive categories. ITA **39**(3), 511–545 (2005)
10. Löwe, M.: Characterisation of parallel independence in AGREE-rewriting. In: Lambers, L., Weber, J. (eds.) ICGT 2018. LNCS, vol. 10887, pp. 118–133. Springer, Cham (2018). https://doi.org/10.1007/978-3-319-92991-0_8
11. Löwe, M.: Double-pushout rewriting in context. In: Mazzara, M., Ober, I., Salaün, G. (eds.) STAF 2018. LNCS, vol. 11176, pp. 447–462. Springer, Cham (2018). https://doi.org/10.1007/978-3-030-04771-9_32
12. Löwe, M.: Double pushout rewriting in context. Technical report 2018/02, FHDW Hannover (2018). www.researchgate.net
13. Robinson, E., Rosolini, G.: Categories of partial maps. Inf. Comput. **79**(2), 95–130 (1988)

Adhesive Subcategories of Functor Categories with Instantiation to Partial Triple Graphs

Jens Kosiol[1]([⊠]) [ID], Lars Fritsche[2] [ID], Andy Schürr[2] [ID], and Gabriele Taentzer[1] [ID]

[1] Philipps-Universität Marburg, Marburg, Germany
{kosiolje,taentzer}@mathematik.uni-marburg.de
[2] TU Darmstadt, Darmstadt, Germany
{lars.fritsche,andy.schuerr}@es.tu-darmstadt.de

Abstract. Synchronization and integration processes of correlated models that are formally based on triple graph grammars often suffer from the fact that elements are unnecessarily deleted and recreated losing information in the process. It has been shown that this undesirable loss of information can be softened by allowing partial correspondence morphisms in triple graphs. We provide a formal framework for this new synchronization process by introducing the category **PTrG** of partial triple graphs and proving it to be adhesive. This allows for ordinary double pushout rewriting of partial triple graphs. To exhibit **PTrG** as an adhesive category, we present a fundamental construction of subcategories of functor categories and show that these are adhesive HLR if the base category already is. Secondly, we consider an instantiation of this framework by triple graphs to illustrate its practical relevance and to have a concrete example at hand.

Keywords: Adhesiveness · Functor category ·
Double pushout rewriting · Triple graphs

1 Introduction

Bidirectional transformation (bx) is a central concept in model-driven software development among others [1,3]. Bx provides the means to define and restore consistency between different kinds of artifacts or different views on a system. Triple Graph Grammars (TGGs) [28] are an established bx-formalism. A triple graph correlates two models (referred to as source and target) by defining a correspondence graph in between that contains elements relating elements of both sides. A TGG defines how correlated models co-evolve and can be used to, e.g., automatically synchronize source and target model after a user edited only one of them. Approaches that have been suggested for such synchronization processes are either informal and rather ad-hoc [9,12], quite inefficient and work under restricted circumstances only [16], or unnecessary deletions may be

© Springer Nature Switzerland AG 2019
E. Guerra and F. Orejas (Eds.): ICGT 2019, LNCS 11629, pp. 38–54, 2019.
https://doi.org/10.1007/978-3-030-23611-3_3

included leading to a loss of information [10,22]. In [7], we present a synchronization process based on triple graphs that allow correspondence morphisms to be partial. This largely improved existing approaches with regard to information loss and runtime. The formal background in that work, however, was restricted and an elaborated theory of partial triple graphs was left to future work. Such a theory is one of the contributions of this paper. We show that the category of partial triple graphs is adhesive such that double pushout rewriting becomes possible [5,21].

When working with adhesive (HLR) categories, an often used technique is to exhibit a category $\mathcal{D}$ as (equivalent to) a functor category $[\mathcal{X}, \mathcal{C}]$ where $\mathcal{X}$ is small and $\mathcal{C}$ is known to be an adhesive (HLR) category. This ensures that $\mathcal{D}$ is an adhesive (HLR) category as well. Considering partial triple graphs and their morphisms, they can be formalized quite naturally as a subcategory of a functor category over an adhesive category. More precisely, they are formalized as those functors of the category $[\bullet \leftarrow \bullet \rightarrow \bullet \leftarrow \bullet \rightarrow \bullet, \mathbf{Graph}]$ that map both central morphisms to injective morphisms. This category can be seen as an instance of a more general principle. There are several categories of interest that form a proper subcategory of a functor category $[\mathcal{X}, \mathcal{C}]$ in a quite natural way. More precisely, we consider subcategories consisting of only those functors that map a designated subset S of the morphisms from $\mathcal{X}$ to monomorphisms (morphisms from $\mathcal{M}$) in $\mathcal{C}$. Besides the already mentioned partial triple graphs, examples include but are not limited to:[1]

1. Elements of $[\bullet \hookrightarrow \bullet, \mathcal{C}]$, where $\mathcal{C}$ is any adhesive (HLR) category, can be understood as objects that come with a marked ($\mathcal{M}$-)subobject. These are exactly the categories that Kastenberg and Rensink proved to be adhesive in [17], for the case that $\mathcal{C}$ is adhesive. They introduce a new concept of attribution for the case where $\mathcal{C}$ is the category **Graph**.
2. Elements of $[\bullet \hookleftarrow \bullet \hookrightarrow \bullet, \mathcal{C}]$, where $\mathcal{C}$ is any adhesive (HLR) category, are exactly the linear rules of $\mathcal{C}$.
3. Elements of $[\bullet \hookleftarrow \bullet \rightarrow \bullet, \mathcal{C}]$ are the partial morphisms of $\mathcal{C}$ (without the usual identification of equivalent ones [27]).
4. Let $\mathcal{S}_n$ be a star-like shape: there exists a central node with n outgoing spans (the shape $\bullet \leftarrow \bullet \rightarrow \bullet \leftarrow \bullet \rightarrow \bullet$ being the special case for $n = 2$). Let $\bar{\mathcal{S}}_n$ denote the same shape with the arrows pointing to the central node designated to be mapped to monomorphisms. König et al. [19,30] use (slice categories of) $[\bar{\mathcal{S}}_n, \mathcal{C}]$ (where $\mathcal{C}$ is a suitable category of models) to formalize a correspondence relation between n different (meta-)models via n partial morphisms from a correspondence model.
5. If $\mathcal{T}$ is a finite tree and $\bar{\mathcal{T}}$ denotes the tree where every edge of $\mathcal{T}$ is marked as to be mapped to monomorphisms (morphisms from $\mathcal{M}$), after fixing an appropriate decoration of the nodes of $\mathcal{T}$ with quantifiers and logical connectives the elements of $[\bar{\mathcal{T}}, \mathcal{C}]$ are nested conditions [13] in $\mathcal{C}$ of a fixed structure.

---

[1] We do not depict identities of $\mathcal{X}$ and mark the morphisms from the designated set S by a hooked arrow.

Simple examples (as Example 4) show that, in general, the full subcategory generated by such a choice of functors will not be an adhesive (HLR) category. In Sect. 4 we show that an adhesive (HLR) category is obtained by restricting the class of morphisms to those natural transformations where all squares induced by the designated morphisms are pullback squares (Theorem 11).

In the second part of this paper (Sect. 5), we instantiate our theory and consider an application to triple graphs with partial morphism between correspondence and source or target graphs. We discuss the expressiveness of double pushout rewriting in that category (Proposition 20) and provide a basic characterization of matches to partial triple graphs (Proposition 21). Moreover, we define the decomposition of a rule on partial triple graphs into a source and a forward rule (Theorem 23) in analogy to the procedure for TGGs [28].

We begin by presenting an introductory example in Sect. 2 and recall some preliminaries in Sect. 3. After the main contributions, we discuss related work in Sect. 6 and conclude in Sect. 7. All omitted proofs are presented in an extended version of this paper [20], together with some additional technical preliminaries.

2   Introductory Example

We motivate our new construction of categories on an example of triple graph grammars (TGGs). TGGs [28] provide a means to define consistency between two correlated models in a declarative and rule-based way by incorporating a third model, a correspondence model, to connect elements from both sides via correspondence links. Elements connected in such a way are henceforth considered to be consistent. Figure 1 shows the rule set of our running example consisting of three TGG rules taken from [7]. They allow to simultaneously create correlated models of a Java abstract syntax tree and a custom documentation model. *Root-Rule* creates a root *Package* and a root *Folder* together with a correspondence link in between which is indicated by the annotation (++) and by green colouring. This rule can be applied arbitrarily often as it does not contain a precondition. *Sub-Rule* requires a *Package*, *Folder*, and a correspondence link between both as precondition and creates a *Package* and *Folder* hierarchy together with a *Doc-File*. Finally, *Leaf-Rule* creates a *Class* with a corresponding *Doc-File* under the same precondition as *Sub-Rule*. Given these rules, we can generate consistent triple graphs like the one depicted in Fig. 2 by iteratively applying the above rules. With *content* contained in the *Doc-Files* on the target side, we indicate that a user has edited them independently such that the model includes information that is private to the target side.

A common use case for TGGs is to synchronize changes between models in order to restore consistency after a user edit. Assume that changes are applied to the left side of Fig. 2 such that the element p is deleted and a new reference is created connecting rootP with subP as depicted in Fig. 3(a). Note that this change also leads to a broken correspondence link, i.e., the result is not a triple graph any longer. There are several TGG-based approaches to synchronize source and target again. However, suggested incremental approaches [9, 12]

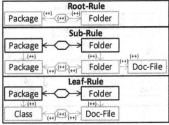

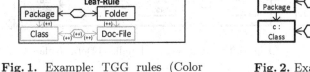

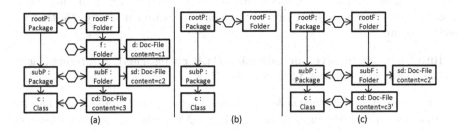

Fig. 1. Example: TGG rules (Color figure online)

Fig. 2. Example: instance model

are rather ad-hoc and not proven to be correct. One provably correct approach parses the whole instance for a maximal remaining valid submodel [16] and is thus quite inefficient and only applicable to a restricted class of TGGs. Other approaches [10,22] first derive the triple graph depicted in Fig. 3(b) and then start a re-translation process using so-called forward rules, which are derived from the TGG, to obtain the consistent triple shown in Fig. 3(c). But the *Doc-Files'* contents have been lost in the process.

Fig. 3. A synchronization scenario

If it was possible to apply rules to partial triple graphs directly, the one depicted in Fig. 3 (a) can be synchronized by applying *Delta-Forward-Rule* as depicted in Fig. 5 where the red elements (additionally annotated with $(--)$) are to be deleted. The qualitative difference of the result is that the contents of both *Doc-Files* are preserved as both elements are not recreated in the process and furthermore, less rule applications are necessary. *Delta-Forward-Rule* can be obtained by splitting *Delta-Rule* (see Fig. 4) into two rules (which is also called *operationalization*): *Delta-Source-Rule* which is a projection to the source component and *Delta-Forward-Rule* which propagates the according changes to correspondence and target graphs. In [8], we show how to construct rules like *Delta-Rule* from given TGG rules. In [7], we operationalize these rules and use them for more efficient synchronization processes. However, we did not introduce partial triple graphs as a category. Moreover, we only defined rule applications to partial triple graphs were the rules arise by operationalizing rules for triple graphs. Hence, an elaborated theory for applying and operationalizing rules for partial triple graphs is still needed which is one of the contributions of this paper.

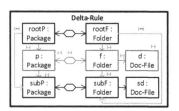

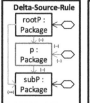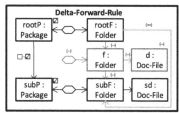

Fig. 4. Example: Delta-Rule (Color figure online)

Fig. 5. Example: Operationalized Delta-Rule (Color figure online)

3 Preliminaries

In this section, we introduce some preliminaries, namely adhesive (HLR) categories and double pushout rewriting. Adhesive categories can be understood as categories where pushouts along monomorphisms behave like pushouts along injective functions in the category of sets. They have been introduced by Lack and Sobociński [21] to offer a unifying formal framework for double pushout rewriting. Later, Ehrig et al. [5] introduced the more general notion of adhesive HLR categories that includes practically relevant examples which are not adhesive. We also introduce the notion of a partial van Kampen square [15] that we need later on.

Definition 1 (Adhesive and adhesive HLR categories). *A category $\mathcal{C}$ with a class of monomorphisms $\mathcal{M}$ is adhesive HLR if*

- *the class of monomorphisms $\mathcal{M}$ contains all isomorphisms and is* closed under composition and decomposition, *i.e., $f, g \in \mathcal{M}$ implies $g \circ f \in \mathcal{M}$ whenever the composition is defined and $g \circ f \in \mathcal{M}, g \in \mathcal{M}$ implies $f \in \mathcal{M}$.*
- *the category $\mathcal{C}$ has pushouts and pullbacks along $\mathcal{M}$-morphisms and $\mathcal{M}$-morphisms are closed under pushouts and pullbacks* such that if Fig. 6 depicts *a pushout square with $m \in \mathcal{M}$ then also $n \in \mathcal{M}$ and analogously if it depicts a pullback square with $n \in \mathcal{M}$ then also $m \in \mathcal{M}$.*
- *pushouts in $\mathcal{C}$ along $\mathcal{M}$-morphisms are* van Kampen squares, *i.e., for any commutative cube as depicted in Fig. 7 where the bottom square is a pushout along an $\mathcal{M}$-morphism m and the backfaces are pullbacks then the top square is a pushout if and only if both front faces are pullbacks.*

A category $\mathcal{C}$ is adhesive if it has all pullbacks, and pushouts along monomorphisms exist and are van Kampen squares.

Pushouts along $\mathcal{M}$-morphisms are partial van Kampen squares if for any commutative cube as depicted in Fig. 7 where the bottom square is a pushout along an $\mathcal{M}$-morphism m, the backfaces are pullbacks, and b and c are $\mathcal{M}$-morphisms, then the top square is a pushout if and only if both front faces are pullbacks and d is an $\mathcal{M}$-morphism.

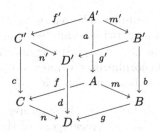

Fig. 6. A pushout square

Fig. 7. Commutative cube over pushout square

Remark 2. Every adhesive category is adhesive HLR for $\mathcal{M}$ being the class of all monomorphisms [5]. Moreover, pushouts along monomorphisms are partial van Kampen squares in adhesive categories [15].

Important examples of adhesive categories include the categories of sets, of (typed) graphs, and of (typed) triple graphs [5,21]. Examples of categories that are not adhesive but adhesive HLR (for an appropriate choice of $\mathcal{M}$) include typed attributed [5] and symbolic attributed graphs [26]. Adhesive HLR categories are a suitable formal framework for rule-based rewriting as defined in the double pushout approach: Rules are a declarative way to define transformations of objects. They consist of a left-hand side (LHS) L, a right-hand side (RHS) R, and a common subobject K, the interface of the rule. In case of (typed) graphs, application of a rule p to a graph G amounts to choosing an image of the rule's LHS L in G, deleting the image of $L\backslash K$ and adding a copy of $R\backslash K$. This procedure can be formalized by two pushouts. Rules and their application semantics are defined as follows.

Definition 3 (Rules and transformations). *Given an adhesive HLR category C, a rule p consists of three objects L, K, and R, called* left-hand side, interface, *and* right-hand side, *and two monomorphisms $l : K \hookrightarrow L, r : K \hookrightarrow R$.*

Given a rule p, an object G, and a monomorphism $m : L \hookrightarrow G$, called match, *a (direct) transformation $G \Rightarrow_{p,m} H$ from G to H via p at match m is given by the diagram to the right where both squares are pushouts.*

$$
\begin{array}{ccccc}
L & \xleftarrow{\ l\ } & K & \xhookrightarrow{\ r\ } & R \\
m\big\uparrow & & \big\uparrow & & \big\uparrow n \\
G & \xleftarrow{\ \ } & D & \xhookrightarrow{\ \ } & H
\end{array}
$$

4   Adhesive Subcategories of Functor Categories

We are interested in investigating subcategories of functor categories $[\mathcal{X}, \mathcal{C}]$ over an adhesive HLR category $\mathcal{C}$ with a set of monomorphisms $\mathcal{M}$. In particular, these subcategories arise by restricting to those functors that map all morphisms from a designated set S of morphisms from $\mathcal{X}$ to $\mathcal{M}$-morphisms in $\mathcal{C}$. As the next example shows, the induced full subcategory fails to be adhesive already for basic examples. The reason for this is that—in the category of morphisms—the (componentwise computed) pushout of monomorphisms does not need to result in a monomorphism again, even if additionally the morphisms between the monomorphisms are monomorphisms as well. The counterexample below has already been presented in [24].

Example 4. Let $\mathcal{C}$ be the adhesive category **Set** and $\mathcal{X}$ the category $\bullet \to \bullet$. Consider the full subcategory of the functor category $[\bullet \to \bullet, \mathbf{Set}]$ induced by those functors that map the only non-identity morphism to an injective function in **Set**. This is just the category with injective functions as objects and commuting squares as morphisms. Let $[n]$ denote the set with n elements and consider the commuting cube depicted to the right: The two squares in the back are a span of monomorphisms in that category. Computing the top and the bottom square as pushouts, i.e., computing the pushout of the two squares in the back in the category $[\bullet \to \bullet, \mathbf{Set}]$,

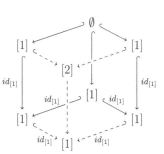

results in a function that is not injective. It is not difficult to check that the category with injective functions as objects and commuting squares as morphisms does not have pushouts, even not along monomorphisms: There is no way to replace the vertical morphism in the front by an injective function and obtain a cube that is a pushout of the two squares in the back.

To resolve this problem, we introduce our categories of interest not as full subcategories of functor categories but restrict the class of allowed morphisms between them.

Definition 5 (S-functor. S-cartesian natural transformation). *Given a small category $\mathcal{X}$, a subset S of the morphisms of $\mathcal{X}$, and an arbitrary category $\mathcal{C}$ with designated class of monomorphisms $\mathcal{M}$, an S-functor is a functor $F : \mathcal{X} \to \mathcal{C}$ such that for every morphism $m \in S$ the morphism Fm is an $\mathcal{M}$-morphism.*

A natural transformation $\sigma : F \to G$ between two S-functors is S-cartesian if for every morphism $S \ni m : x \to y$ the corresponding naturality square $\sigma_y \circ Fm = Gm \circ \sigma_x$ is a pullback square.

Example 6. The partial triple graph that is depicted in Fig. 3(a) is an S-functor from the category $\bullet \leftarrow \bullet \to \bullet \leftarrow \bullet \to \bullet$ to the category of graphs where S consists of the two morphisms to the central object: The left object is mapped to the source graph depicted to the left in Fig. 3(a), the right object to the target graph depicted to the right, and the central object to the correspondence graph consisting of the four hexagons. The second and the fourth object are mapped to the respective domains of the correspondence morphisms to source and target graph. While the domain of the correspondence morphism to the target graph is the whole correspondence graph, the domain of the correspondence morphism to the source graph just consists of three of the hexagons. The outer morphisms of $\bullet \leftarrow \bullet \rightarrow \bullet \leftarrow \bullet \rightarrow \bullet$ are mapped to the correspondence morphisms while the central morphisms are mapped to the inclusion of the domains of the correspondence morphisms into the correspondence graph, which are both injective.

Since the composition of two pullbacks is a pullback and the identity natural transformation is trivially S-cartesian, S-functors with S-cartesian natural transformations as morphisms form a category (associativity of composition and neutrality of the composition with the identity natural transformation are just inherited from the full functor category). We call such categories S-*cartesian subcategories* and denote them by $[\mathcal{X}_S, \mathcal{C}]$.

Proposition 7. *Given a small category $\mathcal{X}$, a subset S of the morphisms of $\mathcal{X}$ and an arbitrary category $\mathcal{C}$ with designated class of monomorphisms $\mathcal{M}$, S-functors with S-cartesian natural transformations as morphisms form a generally non-full subcategory of the functor category $[\mathcal{X}, \mathcal{C}]$. In particular, there is an inclusion functor $I : [\mathcal{X}_S, \mathcal{C}] \hookrightarrow [\mathcal{X}, \mathcal{C}]$.*

Section 5 is devoted to develop the category of partial triple graphs as an instantiation of this very general framework. In particular, it presents concrete examples illustrating the abstract notions. In this section, we prove that if a category $\mathcal{C}$ is adhesive HLR, categories $[\mathcal{X}_S, \mathcal{C}]$ are adhesive HLR again (Theorem 11), assuming pushouts along $\mathcal{M}$-morphisms to be partial van Kampen squares. We first collect results that contribute to the proof of our main theorem. They are of independent interest as they determine that pushouts and pullbacks along $\mathcal{M}$-morphisms are computed componentwise in categories of S-functors. Recall that a functor $F : \mathcal{C} \rightarrow \mathcal{D}$ is said to *create (co-)limits* of a certain type $\mathcal{J}$ if for every diagram $D : \mathcal{J} \rightarrow \mathcal{C}$ and every (co-)limit for $F \circ D$ in $\mathcal{D}$ there exists a unique preimage under F that is a (co-)limit of D in $\mathcal{C}$ [2].

In the following, let $\mathcal{X}$ always be a small category and S a subset of its morphisms; moreover, $\mathcal{C}$ is an arbitrary category with designated class of monomorphisms $\mathcal{M}$ (in particular, if $\mathcal{C}$ is adhesive HLR, $\mathcal{M}$ is understood to be the corresponding class of monomorphisms).

Proposition 8. *Let $[\mathcal{X}_S, \mathcal{C}]$ be an S-cartesian subcategory of $[\mathcal{X}, \mathcal{C}]$. If pullbacks along $\mathcal{M}$-morphisms exist, $[\mathcal{X}, \mathcal{C}]$ and $[\mathcal{X}_S, \mathcal{C}]$ have pullbacks along natural transformations where every component is an $\mathcal{M}$-morphism and the inclusion functor $I : [\mathcal{X}_S, \mathcal{C}] \hookrightarrow [\mathcal{X}, \mathcal{C}]$ creates these.*

The next lemma characterizes the monomorphisms of $[\mathcal{X}_S, \mathcal{C}]$.

Lemma 9. *Let $[\mathcal{X}_S, \mathcal{C}]$ be an S-cartesian subcategory of $[\mathcal{X}, \mathcal{C}]$. Then every morphism in $[\mathcal{X}, \mathcal{C}]$ or in $[\mathcal{X}_S, \mathcal{C}]$ where every component is a monomorphism is a monomorphism in the respective category. If $\mathcal{C}$ has pullbacks, then the converse is true.*

The next proposition states that also pushouts along $\mathcal{M}$-morphisms are calculated componentwise in a category $[\mathcal{X}_S, \mathcal{C}]$. In contrast to the case of pullbacks, the proof requires that $\mathcal{C}$ is adhesive HLR and that pushouts along $\mathcal{M}$-morphisms are partial van Kampen squares.

Proposition 10. *Let C be an adhesive HLR category such that pushouts along $\mathcal{M}$-morphisms are partial van Kampen squares. Let $[\mathcal{X}_S, C]$ be an S-cartesian subcategory of $[\mathcal{X}, C]$. Then $[\mathcal{X}, C]$ and $[\mathcal{X}_S, C]$ have pushouts along morphisms where every component is an $\mathcal{M}$-morphism and the inclusion functor $I : [\mathcal{X}_S, C] \hookrightarrow [\mathcal{X}, C]$ creates these.*

Together, the obtained results guarantee that our construction leads to categories that are adhesive HLR again:

Theorem 11 (Adhesive HLR). *Let C be an adhesive HLR category such that pushouts along $\mathcal{M}$-morphisms are partial van Kampen squares. Let $[\mathcal{X}_S, C]$ be an S-cartesian subcategory of $[\mathcal{X}, C]$. Then $[\mathcal{X}_S, C]$ is adhesive HLR for the class $\mathcal{M}'$ of natural transformations where every component is an $\mathcal{M}$-morphism.*

Proof. By Lemma 9, $\mathcal{M}'$ consists of monomorphisms if C is adhesive HLR and is the class of all monomorphisms if C is adhesive. By definition of $\mathcal{M}'$, the necessary composition and decomposition properties are inherited from $\mathcal{M}$. Since the inclusion functor from $[\mathcal{X}_S, C]$ to $[\mathcal{X}, C]$ creates pullbacks and pushouts along $\mathcal{M}'$-morphisms (Propositions 8 and 10) and $[\mathcal{X}, C]$ is adhesive HLR with respect to natural transformations that are $\mathcal{M}$-morphisms in every component, $[\mathcal{X}_S, C]$ is adhesive HLR as well. $\square$

The next proposition states that—whenever the involved objects and morphisms belong to $[\mathcal{X}_S, C]$—applying a rule in $[\mathcal{X}_S, C]$ or in $[\mathcal{X}, C]$ yields the same result. For simplicity, we suppress the inclusion functor I in its formulation.

Proposition 12 (Functoriality of rule application). *Let C be an adhesive HLR category such that pushouts along $\mathcal{M}$-morphisms are partial van Kampen squares. Let $[\mathcal{X}_S, C]$ be an S-cartesian subcategory of $[\mathcal{X}, C]$. Let $p = (L \xleftarrow{\lambda} K \xrightarrow{\rho} R)$ be a rule and $\mu : L \to G$ a match such that $\lambda, \rho, \mu, L, K, R, G$ are morphisms and objects of $[\mathcal{X}_S, C]$. Then p is applicable to G with match μ in $[\mathcal{X}_S, C]$ if and only if it is in $[\mathcal{X}, C]$. Moreover, the resulting object H coincides (up to isomorphism).*

5   The Category of Partial Triple Graphs

In this section, we apply the theory developed in the section above to the category of (typed) partial triple graphs. Our definition of these rests upon the following (simple) definition of partial morphisms in arbitrary categories. We refrain from identifying equivalent partial morphisms as usually done [27].

Definition 13 (Partial morphism). *A partial morphism $a : A \dashrightarrow B$ is a span $A \xleftarrow{\iota_A} A' \xrightarrow{a} B$ where ι_A is a monomorphism; A' is called the domain of a.*

In the section above, we address the framework of adhesive HLR categories since we generally want to be able to support attribution concepts for partial triple graphs. Two influential such concepts, namely attributed graphs and symbolic attributed graphs, have been shown to constitute adhesive HLR categories [5,26]. Moreover, it is not difficult to check that in both cases pushouts along the respective $\mathcal{M}$-morphisms are partial van Kampen squares. Hence, they can be used as base categories $\mathcal{C}$ when instantiating the framework from above. But for simplicity, we here just present (typed) partial triple graphs without attributes.

Definition 14 (Partial triple graph). *The category of* triple graphs **TrG** *is the functor category* $[\bullet \leftarrow \bullet \rightarrow \bullet, \mathbf{Graphs}]$. *The category of* partial triple graphs **PTrG** *is the category* $[\bullet \leftarrow \bullet \hookrightarrow \bullet \leftarrow \bullet \rightarrow \bullet, \mathbf{Graphs}]$.

Remark 15. By the definitions above, an object $G = (G_S \leftarrow \tilde{G}_S \hookrightarrow G_C \hookleftarrow \tilde{G}_T \rightarrow G_T)$ of **PTrG** might equivalently be considered to consist of a graph G_C with partial morphisms $\sigma : G_C \dashrightarrow G_L$ and $\tau : G_C \dashrightarrow G_T$ where $\tilde{G}_S$ and $\tilde{G}_T$ are the domains of σ and τ, respectively. A morphism $f : G \rightarrow H$ between partial triple graphs then is a triple $(f_S : G_S \rightarrow H_S, f_C : G_C \rightarrow H_C, f_T : G_T \rightarrow H_T)$ of graph morphisms such that both induced squares of partial morphisms commute. In our context, such a square with two opposed partial morphisms (as depicted as square (1) in Fig. 8) *commutes* if there exists a morphism $\tilde{f}_S : \tilde{G}_S \rightarrow \tilde{H}_S$ such that both arising squares (2) and (3) (compare Fig. 9) commute and, moreover, (2) is a pullback square. If $\tilde{f}_S$ exists, it is necessarily unique since ι_{H_S} is a monomorphism. This is stricter than, e.g., the weak commutativity used in [25] that does not require the square (2) to be a pullback square.

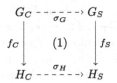

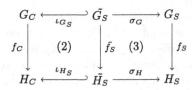

Fig. 8. Square of partial morphisms

Fig. 9. Commuting square of partial morphisms

The next proposition states that the category of triple graphs is isomorphic to a full subcategory of the category of partial triple graphs.

Proposition 16. *The category* **TrG** *of triple graphs is isomorphic to a full subcategory of the category* **PTrG** *of partial triple graphs, i.e., there exists a full and faithful functor* $J : \mathbf{TrG} \rightarrow \mathbf{PTrG}$ *that is injective on objects. Moreover, rule application is functorial.*

In practical applications, the considered triple graphs are generally typed over a fixed triple graph. The next definition introduces typing of partial triple graphs over a fixed triple graph. For, e.g., synchronization scenarios as discussed in [7], it is convenient if the partial triple graphs are still typed over essentially the same triple graph as the original triple graph was.

Definition 17 (Typed partial triple graph). *The category of triple graphs typed over a fixed triple graph $TG = (TG_S \leftarrow TG_C \rightarrow TG_T)$ is the slice category* **TrG**/TG*, denoted* **TrG$_{TG}$**. *The category of partial triple graphs typed over a fixed triple graph TG, denoted* **PTrG$_{TG}$**, *has as objects morphisms $t_G :$ $G \rightarrow J(TG)$ from $[\bullet \leftarrow \bullet \rightarrow \bullet \leftarrow \bullet \rightarrow \bullet,$ **Graphs**] where J is the inclusion functor and G is a partial triple graph. Morphisms are morphisms $g : G \rightarrow H$ from* **PTrG** *such that $t_H \circ g = t_G$.*

Example 18. Figure 3 (a) was already presented as a partial triple graph in Example 6. To consider it as still typed over the same type graph (not depicted) as the original triple graph from Fig. 2, one just restricts its typing morphism accordingly. Note that the resulting typing morphism is not S-cartesian but a morphism in $[\bullet \leftarrow \bullet \rightarrow \bullet \leftarrow \bullet \rightarrow \bullet,$ **Graphs**]. We therefore did not define typing of partial triple graphs just as slice category as well. We exemplify rules and morphisms in **PTrG$_{TG}$** using *Delta-Rule* depicted in Fig. 4: Its LHS consists of the elements depicted in black or in red (not annotated or annotated with $(--)$) while its interface only consists of the black elements. First, the mapping of the interface into the LHS is injective in every component. Moreover, all three correspondence nodes in the LHS of the rule are in the domain of the according correspondence morphism to the source side and the two correspondence nodes that are already part of the interface of the rule are in the domain of the according correspondence morphism as well. This means, the induced square of morphisms is a pullback square. The same holds for the target side. Hence, the morphism from the interface of *Delta-Rule* to its LHS is a morphism in **PTrG$_{TG}$**.

In contrast, the morphism from the interface to the LHS of *Delta-Source-Rule* as depicted in Fig. 5 is not a morphism in **PTrG$_{TG}$**: The according square $\tilde{K}_S \rightarrow K_C \rightarrow L_C \leftarrow \tilde{L}_S \leftarrow \tilde{K}_S$ is not a pullback square since the domain of the correspondence morphism to the source side in the interface graph only contains two elements and not three. Theorem 23 explains why this is not a hindrance to our desired application.

Proposition 19. *Typed partial triple graphs form an adhesive category. Moreover, the category* **TrG$_{TG}$** *of triple graphs typed over TG is isomorphic to a full subcategory of the category* **PTrG$_{TG}$** *of partial triple graphs typed over TG and rule application is functorial.*

We formulate the following results for the category **PTrG**; they hold for categories **PTrG$_{TG}$** as well. The restriction to morphisms where certain squares are required to form pullback squares comes with some limitations. Namely, it is not possible to delete a reference (i.e., an element from the domain of a correspondence morphism) without deleting the referencing element (i.e., the according element in the correspondence graph), nor is it possible to create a reference from an already existing correspondence element. And for every matched correspondence element, a morphism also needs to match the according preimages in the domains of the two correspondence morphisms (if they exist) to become a valid match.

Proposition 20 (Characterizing valid rules and matches). *Let a rule* $p =$ $(L \overset{l}{\hookleftarrow} K \overset{r}{\hookrightarrow} R)$ *and a morphism* $m : L \to G$ *in* $[\bullet \leftarrow \bullet \to \bullet \leftarrow \bullet \to \bullet, \mathbf{Graphs}]$ *be given where* L, K, R, *and* G *are already objects from* **PTrG** *(i.e.,* $L = (L_S \overset{\sigma_L}{\longleftarrow}$ $\tilde{L}_S \overset{\iota_{L_S}}{\longrightarrow} L_C \overset{\iota_{L_T}}{\longleftarrow} \tilde{L}_T \overset{\tau_L}{\longrightarrow} L_T)$ *and similar for* K, R, G*). Then* p *is a rule with match already in* **PTrG** *if and only if (compare Fig. 10 for notation)*

1. $\forall x \in L_C.((x \in \iota_{L_S}(\tilde{L}_S) \wedge x \in l_C(K_C)) \Rightarrow x \in l_C(\iota_{K_S}(\tilde{K}_S)))$ *and analogously* $\forall x \in L_C.((x \in \iota_{L_T}(\tilde{L}_T) \wedge x \in l_C(K_C)) \Rightarrow x \in l_C(\iota_{K_T}(\tilde{K}_T)))$,
2. $\forall x \in R_C.((x \in \iota_{R_S}(\tilde{R}_S) \wedge x \in r_C(K_C)) \Rightarrow x \in r_C(\iota_{K_S}(\tilde{K}_S)))$ *and analogously* $\forall x \in R_C.((x \in \iota_{R_T}(\tilde{R}_T) \wedge x \in r_C(K_C)) \Rightarrow x \in r_C(\iota_{K_T}(\tilde{K}_T)))$, *and*
3. $\forall x \in G_C.((\exists y_1 \in \tilde{G}_S.x = \iota_{G_S}(y_1) \wedge \exists y_2 \in L_C.x = m_C(y_2)) \Rightarrow \exists z \in \tilde{L}_S.(\tilde{m}_S(z) = y_1 \wedge \iota_{L_S}(z) = y_2))$ *and analogously* $\forall x \in G_C.((\exists y_1 \in \tilde{G}_T.x = \iota_{G_T}(y_1) \wedge \exists y_2 \in L_C.x = m_C(y_2)) \Rightarrow \exists z \in \tilde{L}_T.(\tilde{m}_T(z) = y_1 \wedge \iota_{L_T}(z) = y_2))$.

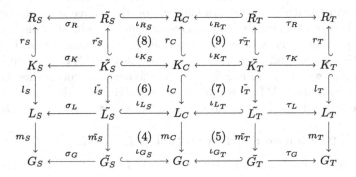

Fig. 10. Rule with match in **PTrG**

We now elementary characterize matches at which a rule in **PTrG** is applicable in the spirit of the gluing condition for graphs [5, Def. 3.9].

Proposition 21. *Let a rule* $p = (L \overset{l}{\hookleftarrow} K \overset{r}{\hookrightarrow} R)$ *and a match* $m : L \to G$ *in* **PTrG** *with* $L = (L_S \overset{\sigma_L}{\longleftarrow} \tilde{L}_S \overset{\iota_{L_S}}{\longrightarrow} L_C \overset{\iota_{L_T}}{\longleftarrow} \tilde{L}_T \overset{\tau_L}{\longrightarrow} L_T)$, $l = (l_S, \tilde{l}_S, l_C, \tilde{l}_T, l_T)$, *and* $m = (m_S, \tilde{m}_S, m_C, \tilde{m}_T, m_T)$ *be given. Then* p *is applicable at match* m *if and only if*

1. m_S, m_C, *and* m_T *satisfy the gluing condition, i.e., none of these morphisms identifies an element that is to be deleted with another element and none of these morphisms determines a node to be deleted that has adjacent edges which are not to be deleted as well [5, Def. 3.9] and*
2. *for every referenced element in the source and the target graphs that is deleted the reference is deleted as well, i.e., for every element* $x \in G_S$ *that has a preimage in* L_S *under* m_S, *no preimage under* $m_S \circ l_S$ *in* K_S *but a preimage in* $\tilde{G}_S$ *under* σ_G, *there is an element* $y \in \tilde{L}_S$ *such that* $m_S(\sigma_L(y)) = x =$

$\sigma_G(\tilde{m}_S(y))$. *Analogously, for every element* $x \in G_T$ *that has a preimage in* L_T *under* m_T, *no preimage under* $m_T \circ l_T$ *in* K_T *but a preimage in* $\tilde{G}_T$ *under* τ_G, *there is an element* $y \in \tilde{L}_T$ *such that* $m_T(\tau_L(y)) = x = \tau_G(\tilde{m}_T(y))$.

Starting point for many practical applications of TGGs is the so-called *operationalization* of a rule which is a split of it into a *source* and a *forward rule* (or equivalently: a target and a backward rule). The source rule only performs the action of the original rule on the source part and the forward rule transfers this to the correspondence and the target part. A basic result states that applying a rule to a triple graph is equivalent to applying the source rule followed by an application of the forward rule. In the rest of this section, we present a comparable definition and result for partial triple graphs. However, we generalize the operationalization of rules in two directions: Our rules are rules on partial triple graphs instead of triple graphs and, moreover, they are allowed to be deleting, whereas classically the rules of a TGG are monotonic [28]. To be able to do this, we need to deviate slightly from the original construction. Our source rules perform the action of the original rule on the source side. Moreover, the deletion-action on the domain of the correspondence morphism to the source part is performed. All other actions are performed by the forward rule. In general, the resulting source and forward rules are not rules in **PTrG** any longer but in $[\bullet \leftarrow \bullet \rightarrow \bullet \leftarrow \bullet \rightarrow \bullet, \mathbf{Graphs}]$. However, the following theorem shows that the application of a rule in **PTrG** is equivalent to applying first the source and afterwards the forward rule (at a suitable match) in $[\bullet \leftarrow \bullet \rightarrow \bullet \leftarrow \bullet \rightarrow \bullet, \mathbf{Graphs}]$.

Definition 22 (Source and forward rule). *Let a rule* $p = (L \overset{l}{\hookleftarrow} K \overset{r}{\hookrightarrow} R)$ *in* **PTrG** *be given. Then its* source rule $p_S = (L^S \overset{l^S}{\hookleftarrow} K^S \overset{r^S}{\hookrightarrow} R^S)$ *is defined as depicted in Fig. 11 where* $\emptyset$ *denotes the empty graph. Its* forward rule $p_F = (L^F \overset{l^F}{\hookleftarrow} K^F \overset{r^F}{\hookrightarrow} R^F)$ *is defined as depicted in Fig. 12.*

$$
\begin{aligned}
L^S &= (L_S \leftarrow \tilde{L}_S \hookrightarrow L_C \hookleftarrow \emptyset \rightarrow \emptyset) \\
&\quad \uparrow l^S \quad \uparrow \quad \uparrow \quad \uparrow \quad \uparrow \quad \uparrow \\
K^S &= (K_S \leftarrow \tilde{K}_S \hookrightarrow L_C \hookleftarrow \emptyset \rightarrow \emptyset) \\
&\quad \downarrow r^S \quad \downarrow \quad \downarrow \quad \downarrow \quad \downarrow \quad \downarrow \\
R^S &= (R_S \leftarrow \tilde{K}_S \hookrightarrow L_C \hookleftarrow \emptyset \rightarrow \emptyset)
\end{aligned}
$$

Fig. 11. Source rule p_S of a rule p

$$
\begin{aligned}
L^F &= (R_S \leftarrow \tilde{K}_S \hookrightarrow L_C \hookleftarrow \tilde{L}_T \rightarrow L_T) \\
&\quad \uparrow l^F \quad \uparrow \quad \uparrow \quad \uparrow \quad \uparrow \quad \cdot \uparrow \\
K^F &= (R_S \leftarrow \tilde{K}_S \hookrightarrow K_C \hookleftarrow \tilde{K}_T \rightarrow K_T) \\
&\quad \downarrow r^F \quad \downarrow \quad \downarrow \quad \downarrow \quad \downarrow \quad \downarrow \\
R^F &= (R_S \leftarrow \tilde{R}_S \hookrightarrow R_C \hookleftarrow \tilde{R}_T \rightarrow R_T)
\end{aligned}
$$

Fig. 12. Forward rule p_F of a rule p

Theorem 23. *Let a rule* $p = (L \overset{l}{\hookleftarrow} K \overset{r}{\hookrightarrow} R)$ *in* **PTrG** *with source and forward rules* $p_S = (L^S \overset{l^S}{\hookleftarrow} K^S \overset{r^S}{\hookrightarrow} R^S)$ *and* $p_F = (L^F \overset{l^F}{\hookleftarrow} K^F \overset{r^F}{\hookrightarrow} R^F)$ *be given.*

1. *Given a direct transformation* $G \Rightarrow_{p,m} H$ *in* **PTrG**, *there is a transformation sequence* $G \Rightarrow_{p_S,m} G' \Rightarrow_{p_F,n} H$ *in* $[\bullet \leftarrow \bullet \rightarrow \bullet \leftarrow \bullet \rightarrow \bullet, \mathbf{Graphs}]$.

2. *Given transformation steps* $G \Rightarrow_{p_S,m} G' \Rightarrow_{p_F,n} H$ *in* $[\bullet \leftarrow \bullet \rightarrow \bullet \leftarrow \bullet \rightarrow$ $\bullet,$ **Graphs**] *where* n *coincides with the comatch of the first transformation step on source and correspondence graph and* G *and* m *are already elements of* **PTrG**, *there is a direct transformation* $G \Rightarrow_{p,m} H$ *in* **PTrG**.

Example 24. Delta-Rule as depicted in Fig. 4 is split by our construction into *Delta-Source-Rule* and *Delta-Forward-Rule* as depicted in Fig. 5. Applying *Delta-Rule* to the triple graph from Fig. 2 such that nodes p and f are deleted gives the same result as first applying *Delta-Source-Rule* at the according match, which gives the partial triple graph from Fig. 3(a) as intermediate result, and *Delta-Forward-Rule* subsequently (with consistent match). Namely, both yield the triple graph from Fig. 3(c).

6 Related Works

In [7] we have already used partial triple graphs to develop an optimized synchronization process for correlated models where the correspondence relationship has been formalized using TGGs. However, we only defined rule application for special cases, only obtained a very restricted version of Theorem 23, and generally left the thorough investigation of that category to future work.

The work that is most closely related to ours with regard to the formal content is the introduction of a new concept of attribution by Kastenberg and Rensink in [17]. They use subgraphs for attribution and prove that the category of reflected monos ($[\bullet \hookrightarrow \bullet, \mathcal{C}]$ in our notation) is adhesive if $\mathcal{C}$ is. Since proving this for an arbitrary S-cartesian subcategory $[\mathcal{X}_S, \mathcal{C}]$ reduces to inspection of a single naturality square at a morphism $m \in S$, our proofs are similar in places. However, they do not consider the case of $\mathcal{C}$ being adhesive HLR and do not relate to the full functor category $[\bullet \rightarrow \bullet, \mathcal{C}]$.

Golas et al. provide a formalization of TGGs in [11] which allows to generalize correspondence relations between source and target graphs as well. They use special typings for the source, target, and correspondence parts of a triple graph and introduce edges between correspondence and source and target nodes instead of using graph morphisms. Hence, they allow for even more flexible correspondence relations and for more flexible deletion and creation of references than possible in our approach. In contrast, we allow for references also between edges and are more in line with the standard formalization of TGGs.

Double pushout rewriting of graph transformation rules by so-called 2-rules has been extensively studied by Machado et al. in [24]. They, too, identify the problem that applying a 2-rule to a rule does not need to result in a rule again since the resulting morphisms are not necessarily injective. Instead of restricting the allowed morphisms, they equip their 2-rules by suitable application conditions. However, their approach is specific to rewriting of graph transformation rules and not directly generalizable to a purely categorical setting. It is not difficult to see that, instantiated for typed graphs, their framework is more general than ours: None of the involved morphisms needs to form a pullback square for a

2-rule to be applicable and result in a rule again. This evokes the research question whether it is possible to increase the classes of morphisms in the categories we presented and still obtain categories that are adhesive HLR.

In contrast to TGGs, where correspondence between elements is defined by total morphisms, partial morphisms have already been used to formalize the correspondence of elements in situations where more than two meta-models are involved [19,30]. As mentioned in Sect. 1, this can also be seen as an instantiation of our general framework. In our practical application to partial triple graphs, we are interested in allowing for partial correspondence morphisms to obtain a more incremental synchronization process. Overall correspondence is still defined via total morphisms.

Partial morphisms have long been a research topic in the area of graph transformation, in particular in connection with the single pushout approach to graph transformation as, e.g., in [4,18,23]. Moreover, there has been research computing limits in categories of partial morphisms [29] or relating properties of pushouts in a category to properties of a pushout in the according category of partial morphisms [14,15]. In this line of research, one enlarges the class of morphisms of a given category by considering also partial morphisms, whereas our framework allows to consider partial morphisms as objects but pushouts and pullbacks are still computed along total morphisms (componentwise).

In [6], Ehrig et al. also consider certain functors as objects of a new category to model distributed objects. They prove (co-)completeness in case the base category is. However, their functors allow for change in the category (or graph) from which the functor starts and the considered morphisms are accordingly quite different from ours.

7   Conclusion

In this paper, we present a new way to construct a category that is adhesive HLR out of a given one, namely as a certain subcategory of a functor category. This construction unifies several categories for which rewriting has been discussed separately so far. As a new application case, we present a category of partial triple graphs. This inspection (as well as comparison to another approach to rewriting rules) shows that, while still interesting in practice, the restriction to a certain kind of morphisms comes with a price. Searching for a (categorical) way to relax this restriction is interesting future work. Moreover, we plan to apply our formal framework to other instances, e.g., to rewriting of constraints.

Acknowledgments. We would like to thank the anonymous reviewers for their valuable feedback. This work was partially funded by the German Research Foundation (DFG), project "Triple Graph Grammars (TGG) 2.0".

References

1. Abou-Saleh, F., Cheney, J., Gibbons, J., McKinna, J., Stevens, P.: Introduction to bidirectional transformations. In: Gibbons, J., Stevens, P. (eds.) Bidirectional Transformations. LNCS, vol. 9715, pp. 1–28. Springer, Cham (2018). https://doi.org/10.1007/978-3-319-79108-1_1
2. Awodey, S.: Category Theory, Oxford Logic Guides, vol. 52, 2nd edn. Oxford University Press Inc., New York (2010)
3. Czarnecki, K., Foster, J.N., Hu, Z., Lämmel, R., Schürr, A., Terwilliger, J.F.: Bidirectional transformations: a cross-discipline perspective. In: Paige, R.F. (ed.) ICMT 2009. LNCS, vol. 5563, pp. 260–283. Springer, Heidelberg (2009). https://doi.org/10.1007/978-3-642-02408-5_19
4. Ehrig, H., et al.: Algebraic approaches to graph transformation - part ii: single pushout approach and comparison with double pushout approach. In: Rozenberg, G. (ed.) Handbook of Graph Grammars and Computing by Graph Transformation, chap. 4, pp. 247–312. World Scientific, Singapore (1997)
5. Ehrig, H., Ehrig, K., Prange, U., Taentzer, G.: Fundamentals of Algebraic Graph Transformation. Monographs in Theoretical Computer Science, Springer, Heidelberg (2006). https://doi.org/10.1007/3-540-31188-2
6. Ehrig, H., Orejas, F., Prange, U.: Categorical foundations of distributed graph transformation. In: Corradini, A., Ehrig, H., Montanari, U., Ribeiro, L., Rozenberg, G. (eds.) ICGT 2006. LNCS, vol. 4178, pp. 215–229. Springer, Heidelberg (2006). https://doi.org/10.1007/11841883_16
7. Fritsche, L., Kosiol, J., Schürr, A., Taentzer, G.: Efficient model synchronization by automatically constructed repair processes. In: Hähnle, R., van der Aalst, W. (eds.) FASE 2019. LNCS, vol. 11424, pp. 116–133. Springer, Cham (2019). https://doi.org/10.1007/978-3-030-16722-6_7
8. Fritsche, L., Kosiol, J., Schürr, A., Taentzer, G.: Short-cut rules. Sequential composition of rules avoiding unnecessary deletions. In: Mazzara, M., Ober, I., Salaün, G. (eds.) STAF 2018. LNCS, vol. 11176, pp. 415–430. Springer, Cham (2018). https://doi.org/10.1007/978-3-030-04771-9_30
9. Giese, H., Hildebrandt, S.: Efficient model synchronization of large-scale models. Technical report 28, Hasso-Plattner-Institut (2009)
10. Giese, H., Wagner, R.: From model transformation to incremental bidirectional model synchronization. Softw. Syst. Modeling 8(1), 21–43 (2009)
11. Golas, U., Lambers, L., Ehrig, H., Giese, H.: Toward bridging the gap between formal foundations and current practice for triple graph grammars. In: Ehrig, H., Engels, G., Kreowski, H.-J., Rozenberg, G. (eds.) ICGT 2012. LNCS, vol. 7562, pp. 141–155. Springer, Heidelberg (2012). https://doi.org/10.1007/978-3-642-33654-6_10
12. Greenyer, J., Pook, S., Rieke, J.: Preventing information loss in incremental model synchronization by reusing elements. In: France, R.B., Kuester, J.M., Bordbar, B., Paige, R.F. (eds.) ECMFA 2011. LNCS, vol. 6698, pp. 144–159. Springer, Heidelberg (2011). https://doi.org/10.1007/978-3-642-21470-7_11
13. Habel, A., Pennemann, K.H.: Correctness of high-level transformation systems relative to nested conditions. Math. Struct. Comput. Sci. **19**, 245–296 (2009)
14. Hayman, J., Heindel, T.: On pushouts of partial maps. In: Giese, H., König, B. (eds.) ICGT 2014. LNCS, vol. 8571, pp. 177–191. Springer, Cham (2014). https://doi.org/10.1007/978-3-319-09108-2_12

15. Heindel, T.: Hereditary pushouts reconsidered. In: Ehrig, H., Rensink, A., Rozenberg, G., Schürr, A. (eds.) ICGT 2010. LNCS, vol. 6372, pp. 250–265. Springer, Heidelberg (2010). https://doi.org/10.1007/978-3-642-15928-2_17

16. Hermann, F., et al.: Model synchronization based on triple graph grammars: correctness, completeness and invertibility. Softw. Syst. Modeling **14**(1), 241–269 (2015)

17. Kastenberg, H., Rensink, A.: Graph attribution through sub-graphs. In: Heckel, R., Taentzer, G. (eds.) Graph Transformation, Specifications, and Nets. LNCS, vol. 10800, pp. 245–265. Springer, Cham (2018). https://doi.org/10.1007/978-3-319-75396-6_14

18. Kennaway, R.: Graph rewriting in some categories of partial morphisms. In: Ehrig, H., Kreowski, H.-J., Rozenberg, G. (eds.) Graph Grammars 1990. LNCS, vol. 532, pp. 490–504. Springer, Heidelberg (1991). https://doi.org/10.1007/BFb0017408

19. König, H., Diskin, Z.: Efficient consistency checking of interrelated models. In: Anjorin, A., Espinoza, H. (eds.) ECMFA 2017. LNCS, vol. 10376, pp. 161–178. Springer, Cham (2017). https://doi.org/10.1007/978-3-319-61482-3_10

20. Kosiol, J., Fritsche, L., Schürr, A., Taentzer, G.: Adhesive subcategories of functor categories with instantiation to partial triple graphs: extended version. Technical report, Philipps-Universität Marburg (2019). https://cms.uni-marburg.de/fb12/arbeitsgruppen/swt/forschung/publikationen/2019/KFST19-TR.pdf/download

21. Lack, S., Sobociński, P.: Adhesive and quasiadhesive categories. Theoret. Inform. Appl. **39**(3), 511–545 (2005)

22. Lauder, M., Anjorin, A., Varró, G., Schürr, A.: Efficient model synchronization with precedence triple graph grammars. In: Ehrig, H., Engels, G., Kreowski, H.-J., Rozenberg, G. (eds.) ICGT 2012. LNCS, vol. 7562, pp. 401–415. Springer, Heidelberg (2012). https://doi.org/10.1007/978-3-642-33654-6_27

23. Löwe, M.: Algebraic approach to single-pushout graph transformation. Theoret. Comput. Sci. **109**(1), 181–224 (1993)

24. Machado, R., Ribeiro, L., Heckel, R.: Rule-based transformation of graph rewriting rules: towards higher-order graph grammars. Theoret. Comput. Sci. **594**, 1–23 (2015)

25. Montanari, U., Ribeiro, L.: Linear ordered graph grammars and their algebraic foundations. In: Corradini, A., Ehrig, H., Kreowski, H.-J., Rozenberg, G. (eds.) ICGT 2002. LNCS, vol. 2505, pp. 317–333. Springer, Heidelberg (2002). https://doi.org/10.1007/3-540-45832-8_24

26. Orejas, F., Lambers, L.: Symbolic attributed graphs for attributed graph transformation. Electronic Communications of the EASST, vol. 30. (International Colloquium on Graph and Model Transformation (GraMoT) 2010) (2010)

27. Robinson, E., Rosolini, G.: Categories of partial maps. Inf. Comput. **79**(2), 95–130 (1988)

28. Schürr, A.: Specification of graph translators with triple graph grammars. In: Mayr, E.W., Schmidt, G., Tinhofer, G. (eds.) WG 1994. LNCS, vol. 903, pp. 151–163. Springer, Heidelberg (1995). https://doi.org/10.1007/3-540-59071-4_45

29. Shir Ali Nasab, A.R., Hosseini, S.N.: Pullback in partial morphism categories. Appl. Categorical Struct. **25**(2), 197–225 (2017)

30. Stünkel, P., König, H., Lamo, Y., Rutle, A.: Multimodel correspondence through inter-model constraints. In: Conference Companion of the 2nd International Conference on Art, Science, and Engineering of Programming, pp. 9–17. ACM, New York (2018)

Extending Predictive Shift-Reduce Parsing to Contextual Hyperedge Replacement Grammars

Frank Drewes[1], Berthold Hoffmann[2], and Mark Minas[3(✉)]

[1] Umeå universitet, Umeå, Sweden
drewes@cs.umu.se
[2] Universität Bremen, Bremen, Germany
hof@uni-bremen.de
[3] Universität der Bundeswehr München, Neubiberg, Germany
mark.minas@unibw.de

Abstract. Parsing with respect to grammars based on hyperedge replacement (HR) is NP-hard in general, even for some fixed grammars. In recent work, we have devised predictive shift-reduce parsing (PSR), a very efficient algorithm that applies to a wide subclass of HR grammars. In this paper, we extend PSR parsing to contextual HR grammars, a moderate extension of HR grammars that have greater generative power, and are therefore better suited for the practical specification of graph and diagram languages. Although the extension requires considerable modifications of the original algorithm, it turns out that the resulting parsers are still very efficient.

Keywords: Graph grammar · Graph parsing ·
Contextual hyperedge replacement

1 Introduction

Grammars based on hyperedge replacement (HR) generate a well-studied class of context-free graph languages [16]. However, their generative power is too weak; e.g., their languages are known to have bounded treewidth [16, Thm. IV.3.12(7)]. Since this even excludes a language as simple as that of all graphs, HR grammars cannot reasonably be advocated for specifying graph models in general.

An example illustrating this weakness of hyperedge replacement is provided by "unstructured" flowcharts with jumps (see Sect. 6). Since jumps can target any location in the program, an edge that represents such a jump may point to any arbitrary node (representing a program location). Inserting such edges is beyond what hyperedge replacement can do because it would require nonterminal hyperedges of unbounded arity, such as the adaptive star grammars of [8].

A similar example is Abstract Meaning Representation [1], a representation of the meaning of natural language sentences that is being heavily studied

© Springer Nature Switzerland AG 2019
E. Guerra and F. Orejas (Eds.): ICGT 2019, LNCS 11629, pp. 55–72, 2019.
https://doi.org/10.1007/978-3-030-23611-3_4

in computational linguistics. Coreferences caused by, e.g., pronouns that refer to entities mentioned elsewhere in a sentence, give rise to edges that point to nodes which may be almost anywhere else in the graph. Hence, again, hyperedge replacement is too weak.

Contextual hyperedge replacement (CHR) has been devised as a moderate extension of HR that overcomes such restrictions, while preserving many other formal properties of HR [7,9]. Rather than having left-hand sides consisting of nonterminal edges only, CHR rules can have additional isolated *context nodes* in their left-hand side, to which the right-hand side can attach edges. Hence, CHR can attach edges to already generated nodes elsewhere in the graph, but the gain in power is limited as the mechanism lacks control over which nodes to choose. (The application conditions of [17] are not yet supported by GRAPPA.)

In recent work, we have devised a very efficient predictive shift-reduce (PSR) parsing algorithm for a subclass of HR grammars. In this paper, we extend this algorithm to contextual HR grammars. Its implementation in the graph-parser distiller GRAPPA[1] turned out to be smooth, and yields parsers that are as efficient as those for the context-free case. This perhaps surprisingly good result is due to the fact that both parsers consume one edge after another and apply rules backwards until the start symbol is reached. As in the context-free case, the grammar analysis by the distiller ensures that suitable edges can be chosen in constant time, and backtracking is avoided. Hence, the overall running time of the generated parser remains linear.

The rest of this paper is structured as follows. Section 2 introduces CHR grammars. In Sect. 3 we recall PSR parsing and discuss the point where it has to be modified for CHR grammars. A particular normal form needed to parse CHR grammars is introduced in Sect. 4, before we discuss the analysis of lookahead in Sect. 5. In Sect. 6 we discuss a more realistic example grammar (of flowcharts), compare the efficiency of different parsers for this grammar, and evaluate some CHR grammars wrt. PSR-parsability with GRAPPA. We conclude by summarizing the results obtained so far, and indicate related and future work in Sect. 7. Due to lack of space, our presentation is driven by a small artificial example, and properties like *unique start nodes* and *free edge choice* are not discussed here in order to keep the paper focused. The complete constructions and proofs for the base case, PSR parsing for HR grammars, can be found in [13].

2   Contextual Hyperedge Replacement Grammars

We first compile the basic notions and notation used in this paper. Throughout the paper, $\mathbb{N}$ denotes the non-negative integers and A^* denotes the set of all finite sequences over a set A, with ε denoting the empty sequence.

We let X be a global, countably infinite supply of *nodes* or *vertices*.

Definition 1 (Graph). An *alphabet* is a set Σ of *symbols* together with an *arity function* $arity\colon \Sigma \to \mathbb{N}$. Then a *literal* $e = \mathsf{a}^{x_1\cdots x_k}$ *over* Σ consists of a

---

[1] Available from its implementor Mark Minas under www.unibw.de/inf2/grappa.

symbol $a \in \Sigma$ and an *attachment* $x_1 \cdots x_k$ of $k = arity(a)$ pairwise distinct nodes $x_1, \ldots, x_k \in X$. We denote the set of all literals over Σ by Lit_Σ.

A *graph* $\gamma = \langle V, \varphi \rangle$ *over* Σ consists of a finite set $V \subseteq X$ of nodes and a sequence $\varphi = e_1 \cdots e_n \in Lit_\Sigma^*$ such that all nodes in these literals are in V. $\mathcal{G}_\Sigma$ denotes the set of all graphs over Σ.

We say that two graphs $\gamma = \langle V, \varphi \rangle$ and $\gamma' = \langle V', \varphi' \rangle$ are *equivalent*, written $\gamma \bowtie \gamma'$, if $V = V'$ and φ is a permutation of φ'.

Note that graphs are sequences of literals, i.e., two graphs $\langle V, \varphi \rangle$ and $\langle V', \varphi' \rangle$ are considered to differ even if $V = V'$ and φ' is just a permutation of φ. However, such graphs are considered equivalent, denoted by the equivalence relation $\bowtie$. "Ordinary" graphs would rather be represented using multisets of literals. The equivalence classes of graphs, therefore, correspond to conventional graphs. The ordering of literals is technically convenient for the constructions in this paper. However, input graphs to be parsed should of course be considered up to equivalence. To make sure that this is the case, our parsers always treat the remaining (not yet consumed) edge literals as a multiset rather than a sequence.

An injective function $\varrho \colon X \to X$ is called a *renaming*. Moreover, γ^ϱ denotes the graph obtained by renaming all nodes in γ according to ϱ.

For a graph $\gamma = \langle V, \varphi \rangle$, we use the notations $X(\gamma) = V$ and $lit(\gamma) = \varphi$. We define the *concatenation* of two graphs $\alpha, \beta \in \mathcal{G}_\Sigma$ as $\alpha\beta = \langle X(\alpha) \cup X(\beta), lit(\alpha)\, lit(\beta) \rangle$. If a graph γ is completely determined by its sequence $lit(\gamma)$ of literals, i.e., if each node in $X(\gamma)$ also occurs in some literal in $lit(\gamma)$, we simply use $lit(\gamma)$ as a shorthand for γ. In particular, a literal $e \in Lit_\Sigma$ is identified with the graph consisting of just this literal and its attached nodes.

We now recall contextual hyperedge replacement from [7,9]. To keep the technicalities simple, we omit node labels. Adding them does not pose any technical or implementational difficulties. Node labels are actually available in GRAPPA, but discussing them here would only complicate the exposition.[2]

Definition 2 (CHR Grammar). Let the alphabet Σ be partitioned into disjoint subsets $\mathcal{N}$ and $\mathcal{T}$ of *nonterminals* and *terminals*, respectively. A *contextual hyperedge replacement rule* $r = (\alpha \to \beta)$ (a *rule* for short) has a graph $\alpha \in \mathcal{G}_\Sigma$ with a single literal $lit(\alpha) = A \in Lit_\mathcal{N}$ as its *left-hand side*, and a graph $\beta \in \mathcal{G}_\Sigma$ with $X(\alpha) \subseteq X(\beta)$ as its *right-hand side*. The nodes in $X(\alpha) \setminus X(A)$ are called *context nodes* of r. A rule without context nodes is called *context-free*.

Consider a graph $\gamma = \delta A' \delta' \in \mathcal{G}_\Sigma$ and a rule r as above. A renaming μ is a *match* (of r to γ) if $A^\mu = A'$ and[3]

$$X(\gamma) \cap X(\beta^\mu) \subseteq X(\alpha^\mu) \subseteq X(\gamma). \tag{1}$$

---

[2] Contextual hyperedge replacement with application conditions, as originally introduced in [17], would require a more significant extension the difficulties of which we have not yet studied. Investigating this will be a topic of future work; cf. Sect. 7.

[3] This condition makes sure that all nodes that are introduced on the right-hand side β of a rule are renamed so that they are distinct from all nodes that do already occur in graph γ, whereas all other nodes are renamed to nodes that occur in γ.

$$Z \Rightarrow_z A^1\underline{B^1} \Rightarrow_{b_2} A^1\underline{D^{13}}b^{13} \Rightarrow_d \underline{A^1}d^{14}d^{43}b^{13} \Rightarrow_a a^{12}b^{13}d^{14}d^{43} = g_1 \qquad (2)$$

$$Z \Rightarrow_z A^1\underline{B^1} \Rightarrow_{b_1} \underline{A^1}C^1b^{13} \Rightarrow_a a^{12}\underline{C^1}b^{13} \Rightarrow_c a^{12}\underline{D^{12}}c^{12}b^{13} \Rightarrow_d a^{12}b^{13}c^{12}d^{14}d^{42} = g_2 \quad (3)$$

$$Z \Rightarrow_z A^1\underline{B^1} \Rightarrow_{b_1} \underline{A^1}C^1b^{13} \Rightarrow_a a^{12}\underline{C^1}b^{13} \Rightarrow_c a^{12}\underline{D^{13}}c^{13}b^{13} \Rightarrow_d a^{12}b^{13}c^{13}d^{14}d^{43} = g_3 \quad (4)$$

Fig. 1. Graphs g_1, g_2, g_3 and their derivations in Γ.

A match μ of r *derives* γ to the graph $\gamma' = \delta\beta^\mu\delta'$. This is denoted as $\gamma \Rightarrow_{r,\mu} \gamma'$, or just as $\gamma \Rightarrow_r \gamma'$. We write $\gamma \Rightarrow_{\mathcal{R}} \gamma'$ for a set $\mathcal{R}$ of rules if $\gamma \Rightarrow_r \gamma'$ for some $r \in \mathcal{R}$, and denote the reflexive-transitive closure of $\Rightarrow_{\mathcal{R}}$ by $\Rightarrow^*_{\mathcal{R}}$, as usual.

A *contextual hyperedge replacement grammar* $\Gamma = (\Sigma, \mathcal{T}, \mathcal{R}, Z)$ (*CHR grammar* for short) consists of finite alphabets $\Sigma, \mathcal{T}$ as above, a finite set $\mathcal{R}$ of rules over Σ, and a *start symbol* $Z \in \mathcal{N}$ of arity 0. Γ generates the language $\mathcal{L}(\Gamma) = \{g \in \mathcal{G}_{\mathcal{T}} \mid Z \Rightarrow^*_{\mathcal{R}} g\}$ of terminal graphs. We call a graph g *valid* with respect to Γ if $\mathcal{L}(\Gamma)$ contains a graph g' with $g \bowtie g'$.

Context-free rules are in fact hyperedge replacement (HR) rules as defined in [13, Def. 2.2], and thus CHR grammars with context-free rules only are HR grammars. Note, however, that derivations in [13] are always *rightmost* derivations that require $\delta' \in \mathcal{G}_{\mathcal{T}}$ in every derivation step. Example 1 demonstrates why derivations for contextual grammars cannot be restricted to just rightmost ones.

Our running example is an artificial CHR grammar chosen for the purpose of illustration only. More practical grammars are considered in Sect. 6.

Example 1. The CHR grammars Γ has $\mathcal{N} = \{Z, A, B, C, D\}$ and $\mathcal{T} = \{a, b, c, d\}$. Z has arity 0, whereas A, B, and C have arity 1; all other labels are binary. Γ has the following rules:

$$Z \xrightarrow{z} A^x B^x \qquad B^x \xrightarrow{b_1 \mid b_2} C^x b^{xy} \mid D^{xy} b^{xy} \qquad D^{xy} \xrightarrow{d} d^{xz} d^{zy}$$

$$A^x \xrightarrow{a} a^{xy} \qquad C^x{+}y \xrightarrow{c} D^{xy} c^{xy}$$

In rule c, $C^x{+}y$ is a shorthand for the graph $\langle\{x, y\}, C^x\rangle$ with context node y; the other rules are context-free.

Figure 1 illustrates the three graphs g_1, g_2, g_3 that constitute the language generated by Γ (up to renaming).[4] As usual, nodes are drawn as circles whereas (binary) edges are drawn as arrows with their label ascribed.

Moreover, Fig. 1 shows the derivations of g_1, g_2, g_3. Underlined nonterminal literals are those rewritten in the next derivation step. The rightmost derivation (2) uses just context-free rules to derive g_1. Both derivation (3) and (4)

---

[4] Thus a CHR grammar is in fact not necessary to describe this tiny language.

use the contextual rule c in their fourth step. There is only a subtle difference: (3) uses node 2 as context node, whereas (4) uses node 3. Neither derivation is rightmost. While (4) could be turned into a rightmost one, there is no rightmost derivation for g_2: In (3), A^1 must be rewritten before C^1 because the rule rewriting C^1 uses 2 as a context node, which has to be created by rewriting A^1.

□

3 Making Shift-Reduce Parsing Predictive

In this section, we recall how a nondeterministic (and inefficient) shift-reduce parsing is made predictive (and efficient), by using a characteristic finite-state automaton (CFA) for control, and by inspecting lookahead. The algorithm developed for HR grammars [13] carries over to CHR grammars in many cases; an exception that requires some modification is discussed at the end of this section.

Nondeterministic Shift-Reduce Parsers read an input graph, and keep a stack of literals (nonterminals and terminals) that have been processed so far. They perform two kinds of actions. *Shift* reads an unread literal of the input graph and pushes it onto the stack. *Reduce* can be applied if literals on top of the stack form the right-hand side of a rule (after a suitable match of their nodes). Then the parser pops these literals off the stack and pushes the left-hand side of the rule onto it (using the same match). The parser starts with an empty stack and an input graph, and accepts this input if the stack just contains the start graph Z and the input has been read completely.

Table 1 shows a parse of graph g_3 in Example 1. Each row shows the current stack, which grows to the right, the sequence of literals read so far, and the multiset of yet unread literals. Note that the literals of the input graph can be shifted in any order; the parser has to choose the literals so that it can construct a reverse rightmost derivation of the input graph. The last column in

Table 1. Shift-reduce parse of graph g_3.

#	Stack	Read literals	Unread literals	Action
0	ε	ε	$\{a^{12}, b^{13}, c^{13}, d^{14}, d^{43}\}$	shift a^{12}
1	$\underline{a^{12}}$	a^{12}	$\{b^{13}, c^{13}, d^{14}, d^{43}\}$	reduce a
2	A^1	a^{12}	$\{b^{13}, c^{13}, d^{14}, d^{43}\}$	shift d^{14}
3	$A^1 d^{14}$	$a^{12} d^{14}$	$\{b^{13}, c^{13}, d^{43}\}$	shift d^{43}
4	$A^1 \underline{d^{14} d^{43}}$	$a^{12} d^{14} d^{43}$	$\{b^{13}, c^{13}\}$	reduce d
5	$A^1 D^{13}$	$a^{12} d^{14} d^{43}$	$\{b^{13}, c^{13}\}$	shift c^{13}
6	$A^1 \underline{D^{13} c^{13}}$	$a^{12} d^{14} d^{43} c^{13}$	$\{b^{13}\}$	reduce c
7	$A^1 C^1$	$a^{12} d^{14} d^{43} c^{13}$	$\{b^{13}\}$	shift b^{13}
8	$A^1 \underline{C^1 b^{13}}$	$a^{12} d^{14} d^{43} c^{13} b^{13}$	$\varnothing$	reduce b_1
9	$\underline{A^1 B^1}$	$a^{12} d^{14} d^{43} c^{13} b^{13}$	$\varnothing$	reduce z
10	Z	$a^{12} d^{14} d^{43} c^{13} b^{13}$	$\varnothing$	accept

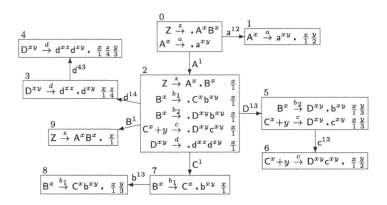

Fig. 2. Characteristic items and states for the steps of the parse of graph g_3 in Table 1.

the table indicates the parser action that yields the situation shown in the next row. Underlined literals on the stack are those popped off the stack by the next reduction step.

We have shown in [13, Sect. 4] that such a parser can find an accepting parse if and only if the input graph is valid. But this parser is highly nondeterministic. For instance, it could start with shifting any literal in step 0, but only shifting a^{12} leads to a successful parse. So it must employ expensive backtracking whenever it runs into a dead end.

Characteristic Finite-State Automata (CFAs) are used to reduce the non-determinism in shift-reduce parsers. This concept has been transferred from LR parsing for context-free string grammars in [13, Sect. 5–7]. The CFA records *items* of the grammar which the parser is processing, where an item is a rule with a dot in its right-hand side that indicates how far processing has advanced. Figure 2 shows the characteristic items for the steps of the parse in Table 1. The numbering of item sets corresponds to the steps in Table 1. Each of these sets corresponds to a state of the CFA, with a particular renaming of nodes.[5]

In step 0 of the parse, the parser starts with the item $Z \xrightarrow{z} .A^x B^x$, where the dot at the beginning indicates that nothing has been processed so far. As the dot is before the nonterminal literal A^x, which can be replaced using rule a, the corresponding configuration in Fig. 2 also contains the item $A^x \xrightarrow{a} .a^{xy}$. So the parser can read a literal a^{xy} (with a suitable match of x and y). It cannot process the nonterminal A^x, as only terminals can be shifted; so shifting a^{12} is the only action fitting the grammar in step 0. As a consequence, step 1 is characterized by the sole item $A^x \xrightarrow{a} a^{xy} . \frac{x}{1} \frac{y}{2}$, which indicates that the right-hand side of rule a has been processed completely and x and y have been matched to nodes 1 and 2, respectively. This implies that a^{12} can be reduced to A^1, which turns item $Z \xrightarrow{z} .A^x B^x$ of step 0 into item $Z \xrightarrow{z} A^x . B^x \frac{x}{1}$ in step 2. Step 2 contains further items because the dot is in front of nonterminal B^x.

---

[5] The complete CFA for Example 1 will only be presented in the next section.

Table 2. Wrong shift-reduce parse of graph g_4. Steps 0–4 are essentially the same as in Table 1

#	Stack	Read literals	Unread literals	Action
0	ε	ε	$\{a^{12}, b^{12}, d^{14}, d^{42}\}$	shift a^{12}
1	$\underline{a^{12}}$	a^{12}	$\{b^{12}, d^{14}, d^{42}\}$	reduce a
2	A^1	a^{12}	$\{b^{12}, d^{14}, d^{42}\}$	shift d^{14}
3	$A^1 d^{14}$	$a^{12}d^{14}$	$\{b^{12}, d^{42}\}$	shift d^{42}
4	$A^1 \underline{d^{14}d^{42}}$	$a^{12}d^{14}d^{42}$	$\{b^{12}\}$	reduce d
5	$A^1 D^{12}$	$a^{12}d^{14}d^{42}$	$\{b^{12}\}$	shift b^{12}
6	$A^1 \underline{D^{12}b^{12}}$	$a^{12}d^{14}d^{42}b^{12}$	$\varnothing$	reduce b_2
7	$\underline{A^1 B^1}$	$a^{12}d^{14}d^{42}b^{12}$	$\varnothing$	reduce z
8	Z	$a^{12}d^{14}d^{42}b^{12}$	$\varnothing$	accept $\wr$

Transitions in the CFA move the dot across a literal in some of its items, and match nodes accordingly. The transitions in Fig. 2 are labeled with these literals. Note that the sequence of literals along any path starting in state 0 equals the stack of the step that is reached by the path.

Every shift-reduce parser can be controled by the CFA so that it will only choose actions in accordance with the grammar, which are promising to find a successful parse if the input graph is valid [13, Sect. 9]. Still some sources of nondeterminism remain, which have to be resolved by different means.

Lookahead may be used when a set of items allows several promising actions. For instance, consider step 5 in Fig. 2, where the dot is in front of b^{13} and c^{13} (under the match $\frac{x}{1}\frac{y}{3}$), which both occur in the unread input. Only shifting c^{13} leads to a successful parse of g_3. If the parser shifted b^{13} instead, the next steps would reduce for rules b_2 and z, yielding a stack Z, leaving c^{13} unread so that the parse would fail. In such a situation, one must check which literals may follow later when either of the actions is selected. An analysis of grammar Γ (prior to parsing) reveals that selecting c^{13} will allow to shift b^{13} later, whereas selecting b^{13} will never allow to shift c^{13} later. So the predictive shift-reduce (PSR) parser must shift c^{13} in step 5. In general (not for Γ), a HR grammar may have states with *conflicts* where the lookahead does not determine which shift or reduction should be done in some state. Then the grammar is not PSR-parsable [13, Sect. 9]. We will discuss the analysis of lookahead in Sect. 5.

Context Nodes require a modification of the PSR parsing algorithm. For a HR grammar, a PSR parser can always continue its parse to a successful one (for *some* remaining input) as long as all actions comply with the CFA. This does not necessarily hold for CHR grammars.

For instance, consider the invalid graph $g_4 = a^{12}d^{14}d^{42}b^{12}$. Its parse, shown in Table 2, starts with the same actions 0–5 as for g_3 in Table 1 and Fig. 2. However, only b^{12} is unread in step 5, which is then shifted. Therefore, the parser will eventually accept g_4 as all literals have been read, although g_4 is invalid! In fact, the reduction in step 6 is wrong, because condition (1) in Definition 2 is violated:

The reduce action is the reverse of the derivation $A^1B^1 \Rightarrow_{b_2} A^1D^{12}b^{12}$ that creates node 2, which is also created when deriving A^1. But the error happened already in step 5 with its characteristic items $B^x \xrightarrow{b_2} D^{xy} \mathbf{.} b^{xy} \frac{x}{1} \frac{y}{2}$ and $C^x{+}y \xrightarrow{c} D^{xy} \mathbf{.} c^{xy} \frac{x}{1} \frac{y}{2}$. Since node 2 has been reused for y, it must be a context node, i.e., the first of these items, which is based on the context-free rule b_2, is not valid in step 5. Apparently, the CFA does not treat context nodes correctly. We shall see in the following section that a CHR grammar like Γ has to be transformed before parsing in order to solve the problem.

4   IOC Normalform

The problem described in the previous section could have been avoided if the parser would "know" in step 5 that node 2 of the nonterminal literal D^{12} is a context node. This would be easy if it held for all occurrences of a D-literal in Γ, but this is not true for the occurrence of D^{xy} in b_2. In the following, we consider only CHR grammars where such situations cannot occur. We require that the label of a nonterminal literal always determines the *roles* of its attached nodes. We then say that a CHR grammar is in *IOC normalform*. Fortunately, every CHR grammar can be transformed into an equivalent IOC normalform. Before we define the IOC normalform in a formal manner, we discuss roles of nodes and profiles of nonterminals.

Nonterminal literals are produced by reduce actions, i.e., if the dot is moved across the nonterminal literal within a rule. For instance, consider literal D^{xy} in item $B^x \to \mathbf{.} D^{xy}b^{xy} \frac{x}{1}$ in Fig. 2. Node x is already bound to node 1 before the dot is moved across D^{xy}, whereas y is unbound before, but bound afterwards (to node 3 in step 5). Nodes x and y act like in and out parameters, respectively, of a procedure; we say that x has role I (for "in") and y has role O (for "out"). By combining I and O for x and y, we say that D^{xy} has *profile IO* in this particular situation. However, the situation is different for D^{xy} in item $C^x{+}y \to \mathbf{.} D^{xy}c^{xy} \frac{x}{1}$. Node x again has role I, and y is again not bound before moving the dot across D^{xy}. But y must then be bound to the context node of this rule; we say that y has role C (for "context"), and D^{xy} has *profile IC* in this situation. So the profile of D^{xy} is not determined by its label D. This must not happen for a CHR grammar in IOC normalform.

Definition 3. A CHR grammar $\Gamma = (\Sigma, \mathcal{T}, \mathcal{R}, \mathsf{Z})$ is in *IOC normalform* if there is a function $P \colon \mathcal{N} \to \{I, O, C\}^*$ so that, for every rule $(\alpha \to \beta) \in \mathcal{R}$ and every nonterminal literal $\boldsymbol{B} = B^{y_1 \cdots y_m} \in Lit_{\mathcal{N}}$ occurring in β, i.e., $\beta = \delta \boldsymbol{B} \delta'$ for some $\delta, \delta' \in Lit_{\Sigma}^*$, $P(B) = p_1 \cdots p_m$ with

$$
p_i = \begin{cases} I & \text{if } y_i \in X(\delta) \\ O & \text{if } y_i \notin X(\delta) \wedge y_i \notin X(\alpha) \\ C & \text{if } y_i \notin X(\delta) \wedge y_i \in X(\alpha) \setminus X(\boldsymbol{A}) \\ p_j' & \text{if } y_i \notin X(\delta) \wedge y_i = x_j \end{cases} \qquad (\text{for } i = 1, \ldots, m)
$$

where $lit(\alpha) = \boldsymbol{A} = A^{x_1 \cdots x_k}$ and $P(A) = p_1' \cdots p_k'$.

Example 2. CHR grammar Γ of Example 1 can be turned into a CHR grammar Γ' in IOC normalform by splitting up the nonterminal label D into D_1 and D_2 and using the following rules:

$$Z \xrightarrow{z} A^x B^x \qquad B^x \xrightarrow{b_1 | b_2} C^x b^{xy} \mid D_1^{xy} b^{xy} \qquad D_1^{xy} \xrightarrow{d_1} d^{xz} d^{zy}$$

$$A^x \xrightarrow{a} a^{xy} \qquad C^x + y \xrightarrow{c} D_2^{xy} c^{xy} \qquad D_2^{xy} \xrightarrow{d_2} d^{xz} d^{zy}$$

It can easily be verified that the function P defined by $Z \mapsto \varepsilon$, $A \mapsto O$, $B \mapsto I$, $C \mapsto I$, $D_1 \mapsto IO$, and $D_2 \mapsto IC$ satisfies the conditions of Definition 3. In particular, every D_1-literal has profile IO, and every D_2-literal has profile IC.

The general construction is straightforward: the simplest method is to create, for every nonterminal label B, all copies $B_{P(B)}$ in which B is indexed with its $3^{arity(B)}$ possible profiles. Each rule for B is thus turned into $3^{arity(B)}$ rules, and the nonterminal literals in the right-hand sides are annotated according to Definition 3. GRAPPA turns this procedure around to avoid the exponential blow-up in most practically relevant cases, as follows. In all rules for a nonterminal label B, assume first that the profile of the left-hand side is $I^{arity(B)}$, and annotate it accordingly. Then annotate the nonterminal labels in all right-hand sides, again following Definition 3. This may give rise to a number of yet unseen annotated nonterminal labels. Create rules for them by copying the respective original rules as before, and repeat until no more new annotations are encountered.

Figure 3 shows the CFA of grammar Γ'. The start state is q_0, indicated by the incoming arc out of nowhere. The CFA has been built in essentially the same

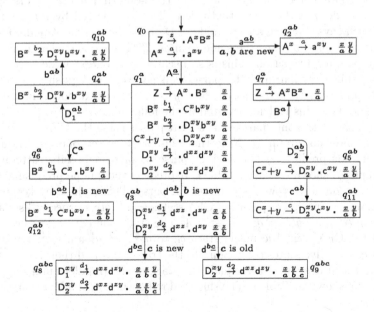

Fig. 3. The characteristic finite automaton of Γ' in Example 2.

way as the characteristic items and states for the steps when parsing the specific graph g_3 in Fig. 2. Instead of the concrete nodes of g_3, we now use *parameters* $\boldsymbol{a}$, $\boldsymbol{b}$, and $\boldsymbol{c}$. They are placeholders, which will be bound to nodes of the particular input graph during parsing. For instance, item $\mathsf{A}^x \xrightarrow{a} \mathsf{a}^{xy} \boldsymbol{\cdot} \frac{x}{1} \frac{y}{2}$ that characterizes step 1 in Fig. 2 corresponds to state $q_2^{\boldsymbol{ab}}$ in Fig. 3 where $\boldsymbol{a}$ and $\boldsymbol{b}$ have been bound to nodes 1 and 2, respectively.

The transitions between states also refer to parameters. For instance, the transition from $q_1^{\boldsymbol{a}}$ to $q_3^{\boldsymbol{ab}}$ means that the dots in the two items $\mathsf{D}_1^{xy} \xrightarrow{d_1} \boldsymbol{\cdot}\, \mathsf{d}^{xz}\mathsf{d}^{zy}\frac{x}{\boldsymbol{a}}$ and $\mathsf{D}_2^{xy} \xrightarrow{d_2} \boldsymbol{\cdot}\, \mathsf{d}^{xz}\mathsf{d}^{zy}\frac{x}{\boldsymbol{a}}$ are moved across d^{xz} where x is bound to $\boldsymbol{a}$ and z is yet unbound. The corresponding shift action must select and read a d-edge in the input graph that leaves the node being bound to parameter $\boldsymbol{a}$. The node of the input graph that matches z and is bound to parameter $\boldsymbol{b}$ in state $q_3^{\boldsymbol{ab}}$, also becomes "known" that way. It must not have been read before because d_1 and d_2 are context-free rules. Figure 3 represents the fact that $\boldsymbol{b}$ is bound to the target node of the d-edge by using $\boldsymbol{b}$ as the underlined target node, and the fact that this node has not been read before is indicated by the label "$\boldsymbol{b}$ is new". Using the IOC normalform, this distinction between "new" and "old" nodes makes it possible to handle context nodes correctly (see also the discussion at the end of Sect. 3). It marks the major technical difference between the context-free parser and the contextual one.

The label "$\boldsymbol{c}$ is old" at the transition from $q_3^{\boldsymbol{ab}}$ to $q_9^{\boldsymbol{abc}}$, however, indicates that $\boldsymbol{c}$ is bound to a node of the input graph that has already been read, together with a shifted edge, earlier in the parsing process. This situation can occur although this transition means moving the dot of item $\mathsf{D}_2^{xy} \xrightarrow{d_2} \mathsf{d}^{xz} \boldsymbol{\cdot}\, \mathsf{d}^{zy}\frac{x}{\boldsymbol{a}}\frac{z}{\boldsymbol{b}}$ across d^{zy} in the context-free rule d_2. Node y of the nonterminal literal corresponds to context node y of the contextual rule c. This is reflected by the profile IC of D_2, which says that the second node of D_2 must be a context node. Further note that $q_8^{\boldsymbol{abc}}$ also contains the same item as $q_9^{\boldsymbol{abc}}$, but $\boldsymbol{c}$ is declared "new" by the corresponding transition. This is so because the context node may be still unread in this situation, and the parser can not yet distinguish whether it is currently processing rule d_2 or d_1, where y is not a context node, indicated by profile IO of D_1. This is demonstrated in the following.

Table 3 shows the only parse that a PSR parser using the CFA in Fig. 3 will try when analyzing graph g_3. Note that this parse corresponds to the unique successful (non-predictive) shift-reduce parse among all possible attempts to parse graph g_3, shown earlier in Table 1. It predicts the unique promising action that keeps it on track towards a successful parse. This is done by keeping, on its stack, an alternating sequence of CFA states and literals processed so far, for instance $q_0\, \mathsf{A}^1\, q_1^1\, \mathsf{d}^{14}\, q_3^{14}\, \mathsf{d}^{43}\, q_8^{143}$ in step 4. The stack contents represent a walk through the CFA from the initial state q_0 to q_8^{143} via q_1^1 and q_3^{14}; the literals between consecutive states correspond to the transitions and their labels. When we ignore the states, the stack equals the stack of the (nondeterministic) shift-reduce parse shown in Table 1. The shift and reduce action of a PSR parser work as follows:

Extending PSR Parsing to Contextual HR Grammars 65

Table 3. PSR parse of g_3 using the CFA in Fig. 3.

#	Stack	Read literals	Unread literals	Action
0	q_0	ε	$\{a^{12},b^{13},c^{13},d^{14},d^{43}\}$	shift a^{12}
1	$q_0\,a^{12}\,q_2^{12}$	a^{12}	$\{b^{13},c^{13},d^{14},d^{43}\}$	reduce a
2	$q_0\,A^1\,q_1^1$	a^{12}	$\{b^{13},c^{13},d^{14},d^{43}\}$	shift d^{14}
3	$q_0\,A^1\,q_1^1\,d^{14}\,q_3^{14}$	$a^{12}d^{14}$	$\{b^{13},c^{13},d^{43}\}$	shift d^{43}
4	$q_0\,A^1\,q_1^1\,d^{14}\,q_3^{14}\,d^{43}\,q_8^{143}$	$a^{12}d^{14}d^{43}$	$\{b^{13},c^{13}\}$	reduce d_2
5	$q_0\,A^1\,q_1^1\,D_2^{13}\,q_5^{13}$	$a^{12}d^{14}d^{43}$	$\{b^{13},c^{13}\}$	shift c^{13}
6	$q_0\,A^1\,q_1^1\,D_2^{13}\,q_5^{13}\,c^{13}\,q_{11}^{13}$	$a^{12}d^{14}d^{43}c^{13}$	$\{b^{13}\}$	reduce c
7	$q_0\,A^1\,q_1^1\,C^1\,q_6^1$	$a^{12}d^{14}d^{43}c^{13}$	$\{b^{13}\}$	shift b^{13}
8	$q_0\,A^1\,q_1^1\,C^1\,q_6^1\,b^{13}\,q_{12}^{13}$	$a^{12}d^{14}d^{43}c^{13}b^{13}$	$\varnothing$	reduce b
9	$q_0\,A^1\,q_1^1\,B^1\,q_7^1$	$a^{12}d^{14}d^{43}c^{13}b^{13}$	$\varnothing$	accept

Table 4. PSR parse of the invalid graph g_4 using the CFA in Fig. 3.

#	Stack	Read literals	Unread literals	Action
0	q_0	ε	$\{a^{12},b^{12},d^{14},d^{42}\}$	shift a^{12}
1	$q_0\,a^{12}\,q_2^{12}$	a^{12}	$\{b^{12},d^{14},d^{42}\}$	reduce a
2	$q_0\,A^1\,q_1^1$	a^{12}	$\{b^{12},d^{14},d^{42}\}$	shift d^{14}
3	$q_0\,A^1\,q_1^1\,d^{14}\,q_3^{14}$	$a^{12}d^{14}$	$\{b^{12},d^{42}\}$	shift d^{42}
4	$q_0\,A^1\,q_1^1\,d^{14}\,q_3^{14}\,d^{42}\,q_9^{142}$	$a^{12}d^{14}d^{42}$	$\{b^{12}\}$	reduce d_2
5	$q_0\,A^1\,q_1^1\,D_2^{12}\,q_5^{12}$	$a^{12}d^{14}d^{42}$	$\{b^{12}\}$	failure

A *shift* action corresponds to an outgoing transition of the state which is currently on top of the stack. For instance, in step 3 with topmost state q_3^{14} there are two transitions leaving q_3^{ab}. They both look for a d-edge leaving node 4 in g_3. The only such edge is d^{43}. And, the parser must choose the transition to q_8^{143} because node 3 is "new", i.e., has not yet occurred in the parse.

A *reduce* action may be selected if the topmost state on the stack contains an item with the dot at the end. For instance, consider step 4 with topmost state q_8^{143}. This state in fact contains two items with a dot at their ends: The parser may either reduce according to rule d_1 or d_2; the CFA cannot help the parser with this decision. However, further analysis (see the following section) reveals that only reducing d_2 can lead to a successful parse. The parser, therefore, removes the topmost four elements from the stack (the right-hand side of rule d_2 together with the states in between, indicated by the underline), leaving q_1^1 as the intermediate topmost state. It then selects the transition for the obtained nonterminal literal D_2^{13} that leaves q_1^1, i.e., the transition to q_5^{13}.

The PSR parser accepts g_3 in step 9 because all literals of g_3 have been read and the topmost state is q_7^1 which has the dot at the end of rule z, i.e., a last reduction would produce Z.

Finally Table 4 shows that the PSR parser using CHR grammar Γ' in IOC normalform correctly recognizes that graph g_4 is invalid. It fails in step 5 in state q_5^{12} where it looks for an unread literal c^{12}, but only finds b^{12}, which cannot be shifted, in contrast to the situation shown in Table 2.

5 Lookahead Analysis

The previous section revealed that the CFA does not always provide enough information for the parser to unambiguously select the next action. This is in fact unavoidable (at least if P $\neq$ NP) because PSR parsing is very efficient while HR graph languages in general can be NP-complete. The problem is that the CFA may contain states with items that trigger different actions, for instance state q_8^{abc} in Fig. 3, which allows two different reduce actions. Then we must analyze (prior to parsing) which literals may follow or cannot follow (immediately or later) in a correct parse when either of the possible actions is chosen. This may yield two results: The analysis may either identify a fixed number of (lookahead) literals that the parser must find among the unread literals in order to always predict the correct next action, or there is at least one state for which this is not possible. The latter situation is called a *conflict*. A CHR grammar is PSR only if no state has a conflict. Here, we describe this situation and the peculiarities for CHR grammars by means of grammar Γ' and state q_8^{abc}.

Consider the situation when the PSR parser has reached q_8^{abc}. We wish to know which literals will be read next or later if either "reduce d_1" or "reduce d_2" is chosen, producing D_1^{ac} or D_2^{ac}. Figure 4 shows the history of the parser that is relevant in this situation. The parser is either in item I_6 or I_7. If it is in I_7, it must also be in I_5 where its literal D_2^{xy} corresponds to the left-hand side of I_7, and so on. B^x in I_1 corresponds to the left-hand side of either I_3 or I_4. Note that the node renamings in $I_2, \ldots, I_5$ reflect the information available when I_6 or I_7 are reduced. For instance, node y of I_5 will be bound to parameter c. However, the choice of context node y of I_5 affects these situations. In this small example, y can be bound either to the node which y of I_2 is bound to, or to the node which y of I_3 is bound to. In the former case, y of I_2 is bound to c since we already know that y of I_5 is bound to c; then, y of I_3 must be a yet unread node, indicated by $\underline{y}$. Otherwise, y of I_3 must be bound to c, and y of I_2 has already been read, but is not stored in any of the parameters, indicated by $\overset{\bullet}{y}$.

It is clear from Fig. 4 that the parser must read a literal c^{ac} next if it reduces I_7, or more precisely, a c-literal that is attached to the input nodes bound to parameters a and c, respectively. This is so because c^{xy} immediately follows D_2^{xy} in I_5 (indicated by a box). And, if the parser reduces I_6, it must read b^{ac} next. In fact, the parser must check whether there is a yet unread literal c^{ac}. If it finds one, it must reduce d_2, otherwise it must reduce d_1. To see this, assume there is an unread c^{ac}. This can be read when I_7 is reduced, but never if I_6 is reduced because no further literal would be read after b^{ac}. And if there is no c^{ac}, the parser would get stuck after reducing I_7.

On the other hand, the parser cannot make a reliable decision based on the existence of just literal b^{ac}, because such a literal can be read by the parser if it

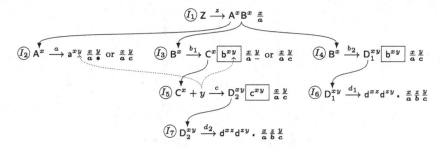

Fig. 4. Lookahead analysis for state q_8^{abc} in Fig. 3.

chooses "reduce d_1", but also if it chooses "reduce d_2". The former is obvious. To see the latter, consider the literal read immediately after c^{ac} when I_7 has been reduced. This must be a literal that corresponds to b^{xy} in I_3, i.e., it is either b^{c-} or b^{ac} according to the possible renamings of I_3. This means, b^{ac} may in fact be read later if I_7 is reduced. Note that this is only possible because the node used as y of I_3 can be the context node used as y of I_5.

6 Realization and Evaluation

PSR parsers for CHR grammars can be generated with the GRAPPA parser distiller (see footnote 1). GRAPPA checks whether the CHR grammar has the *free edge choice property*, which is not discussed in this paper. It ensures that, if the parser can end up in a conflict-like situation between shifting alternative edges, the choice will not affect the outcome of the parsing; see [13, Sect. 9] for details.

Parsing of an input graph starts with the identification of unique start nodes, i.e., a place in the graph where parsing has to start. (This is also not considered in this paper; see [11, Sect. 4] for details.) Then the parser uses the concepts outlined in the previous sections to find a shift-reduce parse of the input graph, and finally tries to construct a derivation from this parse so that context nodes are never used before they have been created in the derivation. However, it may turn out in this last step that the input graph is invalid although a parse has been found. This does happen if there are cyclic dependencies between derivation steps that create nodes and those that use such nodes as context nodes [7]. Identification of start nodes and finding a PSR parse are as efficient as in the context-free case. As discussed in [11,13], this means that these steps require linear time in the size of the graph, for all practical purposes. So does creating a derivation from the parse by topological sorting. As a consequence, PSR parsing with respect to CHR grammars runs in linear time in the size of the input graph. However, a more detailed discussion must be omitted here due to lack of space.

We now use the more realistic language of flowcharts to evaluate PSR parsing for CHR grammars. Note that these "unstructured" flowcharts cannot be specified with HR grammars as they have unbounded treewidth.

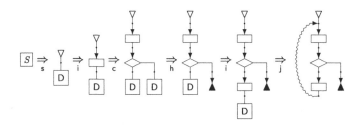

Fig. 5. Derivation of a flowchart

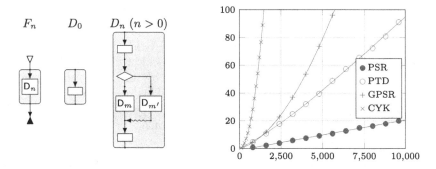

Fig. 6. Definition of flowchart graphs F_n (left) and running time (in milliseconds) of different parsers analyzing F_n for varying numbers of n (right).

Example 3 (Flowcharts). Flowcharts represent the control of low-level imperative programs. In the graphs representing these charts, nodes drawn as small black circles represent program *states*, unary edges labeled with $\triangledown$ and $\blacktriangle$ designate its unique *start state* and its *stop states*, resp., binary edges labeled with $\square$ and $\rightsquigarrow$ represent *instructions* and *jumps*, resp., and ternary edges labeled with $\diamondsuit$ represent *predicates*, which select one of the following branches. (Here, we ignore the texts that usually occur in instructions and predicates.) Flowcharts can be generated by a CHR grammar [7, Ex. 2.6] as follows:

$$S^\varepsilon \xrightarrow{\ \mathsf{s}\ } \triangledown^x \, \mathsf{D}^x \qquad \mathsf{D}^x \xrightarrow{\mathsf{h}|\mathsf{i}|\mathsf{c}} \blacktriangle^x \mid \square^{xy} \, \mathsf{D}^y \mid \diamondsuit^{xyz} \, \mathsf{D}^y \, \mathsf{D}^z \qquad \mathsf{D}^x + y \xrightarrow{\ \mathsf{j}\ } \rightsquigarrow^{xy}$$

The context-free rules h, i, and c generate control flow trees of the halt, instruction, and conditional selection, respectively, and the fourth rule, j, which is not context-free, inserts jumps to some program location in the context. Figure 5 shows the diagrams of a derivation of a small flowchart.

GRAPPA has been used to generate a PSR parser for this grammar. In order to evaluate its efficiency, we have generated three further parsers: a *predictive top-down* (PTD) parser for the same grammar (after extending PTD parsing [10] to CHR grammars), a *Cocke-Younger-Kasami*-style (CYK) parser [20], and finally a *generalized predictive shift-reduce* (GPSR) parser for structured flowcharts [18].

Table 5. Key figures for parsers generated with GRAPPA. "Grammar" columns indicate maximal nonterminal arity (A), nonterminals (N) and terminals (T), context-free (R_{cf}) and contextual (R_c) rules, and the maximal length of right-hand sides (L). As the AMR grammar must be transformed into IOC normalform, its key figures are listed under "IOC normalform". "CFA" columns count states (S), items (I), and transitions (Δ) in the CFAs. The last column counts the conflicts in the CFAs.

Example	Grammar						IOC normalform						CFA			Conflicts
	A	N	T	R_{cf}	R_c	L	A	N	T	R_{cf}	R_c	L	S	I	Δ	
Flowcharts	1	2	5	4	1	3							10	26	30	–
Program graphs	2	12	11	17	4	4							33	80	62	–
AMR	2	9	14	17	11	5	2	13	14	25	11	5	68	211	128	11

GPSR parsing extends PSR parsing to grammars that are not PSR, which is the case for structured flowcharts.

All four parsers have been used to parse flowcharts F_n as defined in Fig. 6 (left), which consist of n predicates, $3n + 1$ instructions, and n jumps. F_n has a subgraph D_n, which, for $n > 0$, contains subgraphs D_m and $D_{m'}$ with $n = m + m' + 1$. Note that the predicates in F_n form a binary tree with n nodes when we ignore instructions. We always choose m and m' such that it is a complete binary tree. Note furthermore that each F_n forms in fact a structured flowchart, which must be built with *jumps* in our flowchart grammar. The GPSR parser has been applied to variations of F_n wherein jumps have been removed, and their source and target nodes have been identified.

Figure 6 shows the running time of each of the four parsers when analyzing graphs F_n with varying values of n. It was measured on a iMac 2017, 4.2 GHz Intel Core i7, Java 1.8.0_202 with standard configuration, and is shown in milliseconds on the y-axis while n is shown on the x-axis. The graphics shows nicely that the PSR parser is linear, and about four times faster than the PTD parser. The GPSR parser is much slower because it deals with conflicts of the CFA by following several parses simultaneously. The CYK parser is the slowest, because it builds up a table of nonterminal edges by dynamic programming.

We have also created parsers for CHR grammars for two additional graph languages: *Program graphs* [7, Ex. 2.7] represent the syntactic structure of object-oriented programming and are used for refactoring. *Abstract Meaning Representations* are widely used in natural language processing to represent sentence meaning. To define the structure of AMRs, CHR grammars are preferable because of their greater capability to cover the set of all AMRs over a given domain. At the same time the grammars become both smaller and simpler. Unfortunately, the example grammar from [14] is not PSR because the CFA has 11 conflicts (see Table 5), but one can employ generalized PSR parsing introduced in [18]. Table 5 lists key figures of the three example languages outlined above.

7 Conclusions

In this paper, we have described how predictive shift-reduce parsing can be extended from HR grammars to CHR grammars. These parsers can be generated with GRAPPA (see footnote 1), and are as efficient as the context-free version, although they apply to a larger class of languages.

Related Work

Much work has been dedicated to graph parsing. Since space is limited, we mention only results and approaches for HR grammars. Early on, Lautemann [19] identified connectedness conditions which make polynomial parsing of certain HR languages possible, using a generalization of the Cocke-Younger-Kasami (CYK) algorithm for context-free string languages. However, the degree of the polynomial depends on the HR language. Stronger connectedness requirements yield cubic parsing (by a different kind of algorithm), as shown by Vogler and Drewes [6,21]. A CYK parser for general HR grammars (even extended by so-called embedding rules) was implemented by Minas in DiaGen [20]. While this parser takes exponential time in general, it can handle graphs with hundreds of nodes and edges.

The line of work continued in this paper started with the proposal of predictive top-down (PTD) parsers in [10] and continued with the introduction of predictive shift-reduce (PSR) parsers [12,13]. Both apply to suitable classes of HR grammars, while the current paper extends PSR parsers to CHR grammars.

Independently, Chiang et al. [4] improved the parser by Lautemann by making use of tree decompositions, and Gilroy et al. [15] studied parsing for the regular graph grammars by Courcelle [5]. Finally, Drewes et al. [2,3] study a structural condition which enables very efficient uniform parsing.

Future Work

If a grammar has conflicts, a *generalized parser* can pursue all conflicting options in parallel until one of them yields a successful parse. This idea, which has been used for LR string parsing in the first place, has recently been transferred to HR grammars [18]. It turns out that generalized PSR parsing can be further extended to CHR grammars. This way the CHR grammar for abstract meaning representations analyzed in Table 5 can be recognized by a generalized PSR parser generated with GRAPPA.

So far, the matching of a context node in a host graph depends on the existence of a matching host node. So a contextual rule may causally depend on another rule that generates the required node. Example 1 showed that certain graphs cannot be derived with a rightmost derivation. This complicates the parser, which always constructs rightmost derivations, since it has to check whether causal dependencies have been respected. Since we conjecture that causal dependencies do not really extend the generative power of CHR grammars, we will consider to re-define CHR grammars, without causality. However, for the practical modeling of graph and diagram languages, it should be possible to express certain conditions that the host node of a context node should fulfill. For instance, one may wish to forbid that the application of a context rule introduces a cycle. This is why the initial version of CHR grammars introduced in [17] features contextual rules with application conditions that can express the existence or absence of certain paths in the host graph. We will investigate the ramifications of application conditions for parsing in the future.

References

1. Banarescu, L., et al.: Abstract meaning representation for sembanking. In: Proceedings of 7th Linguistic Annotation Workshop at ACL 2013 Workshop, pp. 178–186 (2013)
2. Björklund, H., Drewes, F., Ericson, P.: Between a rock and a hard place – uniform parsing for hyperedge replacement DAG grammars. In: Dediu, A.-H., Janoušek, J., Martín-Vide, C., Truthe, B. (eds.) LATA 2016. LNCS, vol. 9618, pp. 521–532. Springer, Cham (2016). https://doi.org/10.1007/978-3-319-30000-9_40
3. Björklund, H., Drewes, F., Ericson, P., Starke, F.: Uniform parsing for hyperedge replacement grammars. UMINF 18.13, Umeå University (2018)
4. Chiang, D., Andreas, J., Bauer, D., Hermann, K.M., Jones, B., Knight, K.: Parsing graphs with hyperedge replacement grammars. In: Proceedings of 51st Annual Meeting of the Association for Computational Linguistic (Vol. 1: Long Papers), pp. 924–932 (2013)
5. Courcelle, B.: The monadic second-order logic of graphs V: on closing the gap between definability and recognizability. Theoret. Comput. Sci. **80**, 153–202 (1991)
6. Drewes, F.: Recognising k-connected hypergraphs in cubic time. Theoret. Comput. Sci. **109**, 83–122 (1993)
7. Drewes, F., Hoffmann, B.: Contextual hyperedge replacement. Acta Informatica **52**, 497–524 (2015)
8. Drewes, F., Hoffmann, B., Janssens, D., Minas, M.: Adaptive star grammars and their languages. Theoret. Comput. Sci. **411**, 3090–3109 (2010)
9. Drewes, F., Hoffmann, B., Minas, M.: Contextual hyperedge replacement. In: Schürr, A., Varró, D., Varró, G. (eds.) AGTIVE 2011. LNCS, vol. 7233, pp. 182–197. Springer, Heidelberg (2012). https://doi.org/10.1007/978-3-642-34176-2_16
10. Drewes, F., Hoffmann, B., Minas, M.: Predictive top-down parsing for hyperedge replacement grammars. In: Parisi-Presicce, F., Westfechtel, B. (eds.) ICGT 2015. LNCS, vol. 9151, pp. 19–34. Springer, Cham (2015). https://doi.org/10.1007/978-3-319-21145-9_2
11. Drewes, F., Hoffmann, B., Minas, M.: Approximating Parikh images for generating deterministic graph parsers. In: Milazzo, P., Varró, D., Wimmer, M. (eds.) STAF 2016. LNCS, vol. 9946, pp. 112–128. Springer, Cham (2016). https://doi.org/10.1007/978-3-319-50230-4_9
12. Drewes, F., Hoffmann, B., Minas, M.: Predictive shift-reduce parsing for hyperedge replacement grammars. In: de Lara, J., Plump, D. (eds.) ICGT 2017. LNCS, vol. 10373, pp. 106–122. Springer, Cham (2017). https://doi.org/10.1007/978-3-319-61470-0_7
13. Drewes, F., Hoffmann, B., Minas, M.: Formalization and correctness of predictive shift-reduce parsers for graph grammars based on hyperedge replacement. J. Logical Algebraic Methods Program. **104**, 303–341 (2019)

14. Drewes, F., Jonsson, A.: Contextual hyperedge replacement grammars for abstract meaning representations. In: 13th International Workshop on Tree-Adjoining Grammar and Related Formalisms (TAG+13), pp. 102–111 (2017)
15. Gilroy, S., Lopez, A., Maneth, S.: Parsing graphs with regular graph grammars. In: Proceedings of 6th Joint Conference on Lexical and Computational Semantics (*SEM 2017), pp. 199–208 (2017)
16. Habel, A.: Hyperedge Replacement: Grammars and Languages. LNCS, vol. 643. Springer, Heidelberg (1992). https://doi.org/10.1007/BFb0013875
17. Hoffmann, B., Minas, M.: Defining models - meta models versus graph grammars. Elect. Commun. EASST **29** (2010). Proceedings of 6th Workshop on Graph Transformation and Visual Modeling Techniques (GT-VMT'10), Paphos, Cyprus
18. Hoffmann, B., Minas, M.: Generalized predictive shift-reduce parsing for hyperedge replacement graph grammars. In: Martín-Vide, C., Okhotin, A., Shapira, D. (eds.) LATA 2019. LNCS, vol. 11417, pp. 233–245. Springer, Cham (2019). https://doi.org/10.1007/978-3-030-13435-8_17
19. Lautemann, C.: The complexity of graph languages generated by hyperedge replacement. Acta Informatica **27**, 399–421 (1990)
20. Minas, M.: Concepts and realization of a diagram editor generator based on hypergraph transformation. Sci. Comput. Program. **44**(2), 157–180 (2002)
21. Vogler, W.: Recognizing edge replacement graph languages in cubic time. In: Ehrig, H., Kreowski, H.-J., Rozenberg, G. (eds.) Graph Grammars 1990. LNCS, vol. 532, pp. 676–687. Springer, Heidelberg (1991). https://doi.org/10.1007/BFb0017421

Analysis and Verification

Exploring Conflict Reasons for Graph Transformation Systems

Leen Lambers[1]([✉])(iD), Jens Kosiol[2](iD), Daniel Strüber[3](iD),
and Gabriele Taentzer[2](iD)

[1] Hasso-Plattner-Institut, Universität Potsdam, Potsdam, Germany
leen.lambers@hpi.de
[2] Philipps-Universität Marburg, Marburg, Germany
{kosiolje,taentzer}@informatik.uni-marburg.de
[3] Chalmers University, University of Gothenburg, Gothenburg, Sweden
danstru@chalmers.se

Abstract. Conflict and dependency analysis (CDA) is a static analysis for the detection of conflicting and dependent rule applications in a graph transformation system. Recently, granularity levels for conflicts and dependencies have been investigated focussing on delete-use conflicts and produce-use dependencies. A central notion for granularity considerations are (minimal) conflict and dependency reasons. For a rule pair, where the second rule is non-deleting, it is well-understood based on corresponding constructive characterizations how to efficiently compute (minimal) conflict and dependency reasons. We further explore the notion of (minimal) conflict reason for the general case where the second rule of a rule pair may be deleting as well. We present new constructive characterizations of (minimal) conflict reasons distinguishing delete-read from delete-delete reasons. Based on these constructive characterizations we propose a procedure for computing (minimal) conflict reasons and we show that it is sound and complete.

Keywords: Graph transformation · Conflict analysis · Static analysis

1 Introduction

Graph transformation [1] is a formal paradigm with many applications. A graph transformation system is a collection of graph transformation rules that, in union, serve a common purpose. For many applications (see [2] for a survey involving 25 papers), it is beneficial to know all conflicts and dependencies that can occur for a given pair of rules. A conflict is a situation in which one rule application renders another rule application inapplicable. A dependency is a situation in which one rule application needs to be performed such that another rule application becomes possible. For a given rule set, a *conflict and dependency analysis* (CDA) technique is a means to compute a list of all pairwise conflicts and dependencies.

Inspired by the related concept from term rewriting, *critical pair analysis* (CPA, [3]) has been the established CDA technique for over two decades. CPA

© Springer Nature Switzerland AG 2019
E. Guerra and F. Orejas (Eds.): ICGT 2019, LNCS 11629, pp. 75–92, 2019.
https://doi.org/10.1007/978-3-030-23611-3_5

reports each conflict as a critical pair[1], that is, a minimal example graph together with a pair of rule applications from which a conflict arises. Recently it has been observed that applying CPA in practice is problematic: First, computing the critical pairs does not scale to large rules and rule sets. Second, the results are often hard to understand; they may contain many redundant conflicts that differ in subtle details only, typically not relevant for the use case at hand.

To address these drawbacks, in previous work, we presented the *multi-granular conflict and dependency analysis* (MultiCDA [2,5]) for graph transformation systems. It supports the computation of conflicts on a given granularity level: On binary granularity, it reports if a given rule pair contains a conflict at all. On coarse granularity, it reports *minimal conflict reasons*, that is, problematic rule fragments shared by both rules that may give rise to a conflict. Fine granularity is, roughly speaking, the level of granularity provided by critical pairs. (In the terminology of our recent work [4], we here focus on different levels of overlap granularity with fixed coarse context granularity.) We showed that coarse-grained results are more usable than fine-grained ones in a diverse set of scenarios and can be used to compute the fine-grained results much faster.

In this work, we address a major *current limitation of MultiCDA* [2]. The computation of conflicts is only exact for cases where the second rule of the considered rule pair is non-deleting. In this case, it is well-understood how to compute conflicts efficiently, using constructive characterisations of minimal conflict reasons (for coarse granularity) and conflict reasons (for fine granularity). In the other case, MultiCDA can provide an overapproximation of the actual conflicts, by replacing the second rule with its non-deleting variant. On the one hand, this overapproximation may contain conflicts which can never actually arise. On the other hand, the overapproximation does not distinguish conflicts for rule pairs where the second rule is non-deleting from those for rule pairs where the second rule is deleting (i.e. no distinction between delete-delete and delete-read). The first issue leads to a MultiCDA that may report false positives, whereas the latter issue causes the MultiCDA to report true positives without the desired level of detail. Therefore the overapproximation presents an obstacle to the understandability of the results and the usability of the overall technique.

In this paper, we come up with the foundations for an *improved MultiCDA* avoiding this limitation, thus delivering in all cases exact as well as detailed results. To this end we present new constructive characterizations of (minimal) conflict reasons for rule pairs where the second rule may be deleting, distinguishing in particular delete-read (dr) from delete-delete (dd) reasons. Based on these constructive characterizations we propose a basic procedure for computing dr/dd (minimal) conflict reasons and we show that it is sound and complete. In particular, we learn that we can reduce the computation of dr/dd minimal reasons to the constructions presented for the overapproximation [2]. Moreover, the construction of dr/dd reasons can reuse the results computed for minimal reasons, representing the basis for an efficient computation of fine-grained results

---

[1] For brevity, since all conflict-specific considerations in this paper dually hold for dependencies (see our argumentation in [4]), we omit talking about dependencies.

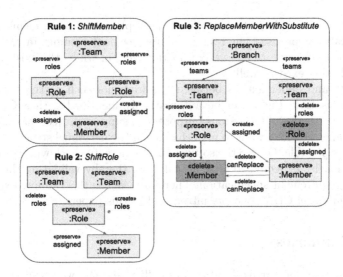

Fig. 1. Rules for running example in an integrated representation

based on coarse-grained ones. We illustrate our results using a running example modeling requirements for a project management software.

The rest of this paper is structured as follows: Sect. 2 introduces our running example. Section 3 revisits preliminaries. Section 4 introduces *delete-read* and *delete-delete* conflict reasons. Section 5 presents a new characterization of conflict reasons, accommodating both delete-read (dr) and delete-delete (dd) conflicts. Section 6 is devoted to the construction of conflict reasons based on the new characterizations. Section 7 discusses related work and concludes. We present proofs that are omitted from this paper in an extended version [6].

2   Running Example

In agile software development processes [7], enterprises quickly react to changes by flexibly adapting their team structures. Figure 1 introduces a set of rules describing requirements for a project management software. The rules are represented in the Henshin [8] syntax, using an integrated syntax with delete, create, and preserve elements. Delete and create elements are only contained in the LHS and RHS, respectively, whereas a preserved element represents an element that occurs both in the LHS and RHS.

The rules, focusing on restructuring and deletion cases for illustration purposes, stem from a larger rule overall set. The first two rules allow the project managers of a branch of the company to reassign roles and members between teams. Rule *ShiftMember* assigns a member to a different role in the same team. Rule *ShiftRole* moves a role and its assigned team member to a different team. The other rule deals with a team member leaving the company. Rule *Replace-MemberWithSubstitute* (abbreviated to *ReplaceM* in what follows) removes a team member while filling the left role with a replacement team member, based on their shared expertise. The role of the replacement member in the existing project is deleted. Note that a variant of this rule, in which the existing role is not deleted, may exist as well.

Conflicts and dependencies between requirements such as those expressed with the rules from Fig. 1 can be automatically identified with a CDA technique. Doing so is useful for various purposes: To support project managers from different teams who may want to plan changes to the personnel structure as independently as possible. Or, for the software developers, to check whether conflicts and dependencies expected to arise actually do, thereby validating the correctness of the requirement specification. Therefore, we use this running example to illustrate the novel CDA concepts introduced in this paper.

3   Preliminaries

As a prerequisite for our exploration of conflict reasons, we recall the double-pushout approach to graph transformation as presented in [1]. Furthermore, we reconsider conflict notions on the transformation and rule level [2,9,10], including conflict reasons.

3.1   Graph Transformation and Conflicts

Throughout this paper we consider graphs and graph morphisms as presented in [1] for the category of graphs; all results can be easily extended to the category of typed graphs by assuming that each graph and morphism is typed over some fixed type graph TG.

Graph transformation is the rule-based modification of graphs. A *rule* mainly consists of two graphs: L is the left-hand side (LHS) of the rule representing a pattern that has to be found to apply the rule. After the rule application, a pattern equal to R, the right-hand side (RHS), has been created. The intersection K is the graph part that is not changed; it is equal to $L \cap R$ provided that the result is a graph again. The graph part that is to be deleted is defined by $L \setminus (L \cap R)$, while $R \setminus (L \cap R)$ defines the graph part to be created.

A *direct graph transformation* $G \overset{m,r}{\Longrightarrow} H$ between two graphs G and H is defined by first finding a graph morphism[2] m of the LHS L of rule r into G such that m is injective, and second by constructing H in two passes: (1) build $D := G \setminus m(L \setminus K)$, i.e., erase all graph elements that are to be deleted; (2) construct $H := D \cup m'(R \setminus K)$. The morphism m' has to be chosen such that a new copy of all graph elements that are to be created is added. It has been shown for graphs and graph transformations that r is applicable at m iff m fulfills the *dangling condition*. It is satisfied if all adjacent graph edges of a graph node to be

---

[2] A morphism between two graphs consists of two mappings between their nodes and edges being both structure-preserving w.r.t. source and target functions. Note that in the main text we denote inclusions by $\hookrightarrow$ and all other morphisms by $\rightarrow$.

deleted are deleted as well, such that D becomes a graph. Injective matches are usually sufficient in applications and w.r.t. our work here, they allow to explain constructions with more ease than for general matches. In categorical terms, a direct transformation step is defined using a so-called double pushout as in the following definition. Thereby step (1) in the previous informal explanation is represented by the first pushout and step (2) by the second one [1].

Definition 1 ((non-deleting) rule and transformation). *A rule r is defined by $r = (L \overset{le}{\hookleftarrow} K \overset{ri}{\hookrightarrow} R)$ with L, K, and R being graphs connected by two graph inclusions. The non-deleting rule of r is defined by $ND(r) = (L \overset{id_L}{\hookleftarrow} L \overset{ri'}{\hookrightarrow} R')$ with $(L \overset{ri'}{\hookrightarrow} R' \hookleftarrow R)$ being the pushout of (le, ri). Given rule $r_1 = (L_1 \overset{le_1}{\hookleftarrow} K_1 \overset{ri_1}{\hookrightarrow} R_1)$, square (1) in Fig. 2 can be constructed as initial pushout over morphism le_1. It yields the* boundary graph B_1 *and the* deletion graph C_1.

A direct transformation $G \overset{m,r}{\Longrightarrow} H$ *which applies rule r to a graph G consists of two pushouts as depicted right. Rule r is* applicable *and the injective morphism $m : L \to G$ is called* match *if there exists a graph D such that $(PO1)$ is a pushout.*

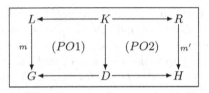

Given a pair of transformations, a *delete-use conflict* [1] occurs if the match of the second transformation cannot be found anymore after applying the first transformation. Note that we do not consider delete-use conflicts of the second transformation on the first one explicitly. To get those ones as well, we simply consider the inverse pair of transformations. The following definition moreover distinguishes two cases [9]: (1) a *delete-read conflict* occurs if the match of the first transformation can still be found after applying the second one (2) a *delete-delete conflict* occurs if this is not the case, respectively.

Definition 2 (dr/dd conflict). *Given a pair of direct transformations (t_1, t_2) $= (G \overset{m_1, r_1}{\Longrightarrow} H_1, G \overset{m_2, r_2}{\Longrightarrow} H_2)$ applying rules $r_1 : L_1 \overset{le_1}{\hookleftarrow} K_1 \overset{ri_1}{\hookrightarrow} R_1$ and $r_2 : L_2 \overset{le_2}{\hookleftarrow} K_2 \overset{ri_2}{\hookrightarrow} R_2$ as depicted in Fig. 2. Transformation pair (t_1, t_2) is in* delete-use conflict *if there does not exist a morphism $x_{21} : L_2 \to D_1$ such that $g_1 \circ x_{21} = m_2$. Transformation pair (t_1, t_2) is in* dr conflict *if it is in delete-use conflict and if there exists a morphism $x_{12} : L_1 \to D_2$ such that $g_2 \circ x_{12} = m_1$. Transformation pair (t_1, t_2) is in* dd conflict *if it is in delete-use conflict and if there does not exist a morphism $x_{12} : L_1 \to D_2$ such that $g_2 \circ x_{12} = m_1$.*

3.2 Conflict Reasons

We consider delete-use conflicts between transformations where at least one deleted element of the first transformation is overlapped with some used element of the second transformation. This *overlap* is formally expressed by a span

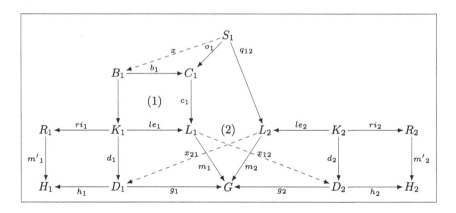

Fig. 2. Illustration of conflict and conflict reason

of graph morphisms between the deletion graph C_1 of the first rule, and the LHS of the second rule (Fig. 2). Remember that $C_1 := L_1 \setminus (K_1 \setminus B_1)$ contains the deletion part of a given rule and boundary graph B_1 consisting of all nodes needed to make $L_1 \setminus K_1$ a graph. $C_1 \setminus B_1$ may consist of several disjoint fragments, called *deletion fragments*. Completing a deletion fragment to a graph by adding all incident nodes (i.e. boundary nodes) it becomes a *deletion component* in C_1. Each two deletion components overlap in boundary nodes only; the union of all deletion components is C_1. If two transformations overlap such that there is at least one element of a deletion fragment included, they are in conflict.

The *overlap conditions* reintroduced in Definition 3 describe for an overlap of a given pair of rules under which conditions it may lead to a conflict (conflict condition), since there exist transformations (transformation condition) that overlap all elements as prescribed by the given overlap indeed (completeness condition). We call such an overlap *conflict reason* and it is minimal if no bigger one exists in which it can be embedded. Table 1 provides an overview over all conflict notions for rules (as reintroduced in Definition 4) and their overlap conditions.

General Setting: For the rest of this paper, we assume the following basic setting: Given rules $r_1 : L_1 \overset{le_1}{\hookleftarrow} K_1 \overset{ri_1}{\hookrightarrow} R_1$ with the initial pushout (1) for $K_1 \overset{le_1}{\hookleftarrow} L_1$ and $r_2 : L_2 \overset{le_2}{\hookleftarrow} K_2 \overset{ri_2}{\hookrightarrow} R_2$, we consider a span $s_1 : C_1 \overset{o_1}{\hookleftarrow} S_1 \overset{q_{12}}{\hookrightarrow} L_2$ as depicted in Fig. 2.

Definition 3 (overlap conditions). *Given rules r_1 and r_2 as well as a span s_1, overlap conditions for the span s_1 of (r_1, r_2) are defined as follows:*

1. Weak conflict condition: *Span s_1 satisfies the* weak conflict condition *if there does not exist any injective morphism $x : S_1 \to B_1$ such that $b_1 \circ x = o_1$.*
2. Conflict condition: *Span s_1 satisfies the* conflict condition *if for each coproduct $\bigoplus_{i \in I} S_1^i$, where each S_1^i is non-empty and $S_1 = \bigoplus_{i \in I} S_1^i$, each of the induced*

spans $s_1^i : C_1 \overset{o_1^i}{\hookleftarrow} S_1^i \overset{q_{12}^i}{\rightarrowtail} L_2$ with $o_1^i = o_1|_{S_1^i}$ and $q_{12}^i = q_{12}|_{S_1^i}$ fulfills the weak conflict condition.

3. Transformation condition: Span s_1 satisfies the transformation condition if there is a pair of transformations $(t_1, t_2) = (G \overset{m_1, r_1}{\Longrightarrow} H_1, G \overset{m_2, r_2}{\Longrightarrow} H_2)$ via (r_1, r_2) with $m_1(c_1(o_1(S_1))) = m_2(q_{12}(S_1))$ (i.e. (2) is commuting in Fig. 2).

4. Completeness condition: Span s_1 satisfies the completeness condition if there is a pair of transformations $(t_1, t_2) = (G \overset{m_1, r_1}{\Longrightarrow} H_1, G \overset{m_2, r_2}{\Longrightarrow} H_2)$ via (r_1, r_2) such that (2) is the pullback of $(m_1 \circ c_1, m_2)$ in Fig. 2.

5. Minimality condition: A span $s'_1 : C_1 \overset{o'_1}{\hookleftarrow} S'_1 \overset{q'_{12}}{\rightarrowtail} L_2$ can be embedded into span s_1 if there is an injective morphism $e : S'_1 \to S_1$, called embedding morphism, such that $o_1 \circ e = o'_1$ and $q_{12} \circ e = q'_{12}$. If e is an isomorphism, then we say that the spans s_1 and s'_1 are isomorphic. (See (3) and (4) in Fig. 3.) Span s_1 satisfies the minimality condition w.r.t. a set SP of spans if any $s'_1 \in SP$ that can be embedded into s_1 is isomorphic to s_1.

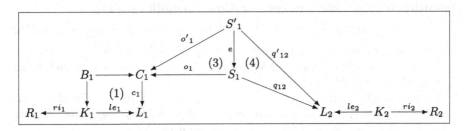

Fig. 3. Illustrating span embeddings

Note that span s_1 which fulfils the weak conflict condition, also fulfils the conflict condition iff S_1 does not contain any isolated boundary nodes [4].

Definition 4 (conflict notions). Let the rules r_1 and r_2 as well as a span s_1 be given.

1. Span s_1 is called conflict part candidate for the pair of rules (r_1, r_2) if it satisfies the conflict condition. Graph S_1 is called the conflict graph of s_1.
2. A conflict part candidate s_1 for (r_1, r_2) is a conflict part for (r_1, r_2) if s_1 fulfils the transformation condition.
3. A conflict part candidate s_1 for (r_1, r_2) is a conflict atom candidate for (r_1, r_2) if it fulfils the minimality condition w.r.t. the set of all conflict part candidates for (r_1, r_2).
4. A conflict part s_1 for (r_1, r_2) is a conflict atom if it fulfils the minimality condition w.r.t. the set of all conflict parts for (r_1, r_2).
5. A conflict part s_1 for (r_1, r_2) is a conflict reason for (r_1, r_2) if s_1 fulfils the completeness condition.

Table 1. Overview of conflict notions

Overlap condition/ conflict notion	Conflict condition	Transf. condition	Compl. condition	Minimality condition
Conflict part candidate	x			
Conflict part	x	x		
Conflict atom candidate	x			x
Conflict atom	x	x		x
Conflict reason	x	x	x	
Min. conflict reason	x	x	x	x

6. *A conflict reason s_1 for (r_1, r_2) is* minimal *if it fulfils the minimality condition w.r.t. the set of all conflict reasons for (r_1, r_2).*

Conflict notions are in various interrelations as shown in [4]. Here, we recall those that are relevant for our further exploration of conflict reasons.

Definition 5 (covering and composition of conflict parts).

1. *Given a conflict part s_1, the set A of all conflict atoms that can be embedded into s_1* covers *s_1 if for each conflict part $s'_1 : C_1 \overset{o'_1}{\hookleftarrow} S'_1 \overset{q'_{12}}{\hookrightarrow} L_2$ for (r_1, r_2) that can be embedded into s_1, it holds that s'_1 is isomorphic to s_1 if each atom in A can be embedded into s'_1.*
2. *Given a conflict part s_1, the set $M = \{s_i^m |\ i \in I\}$ of spans that can be embedded into s_1 via a corresponding set of embedding morphisms $E_M = \{e_i |\ i \in I\}$* composes *s_1 if the set E_M is jointly surjective.*

Fact 1 (Interrelations of conflict notions and characterization [2,4]).
Let rules r_1 and r_2 as well as conflict part candidate s_1 for (r_1, r_2) be given.

1. *If s_1 is a conflict part for (r_1, r_2), there is a conflict reason for (r_1, r_2) such that s_1 can be embedded into it.*
2. *If s_1 is a conflict atom candidate for rules (r_1, r_2), its conflict graph S_1 either consists of a node v s.t. $o_1(v) \in C_1 \setminus B_1$ or of an edge e with its incident nodes v_1 and v_2 s.t. $o_1(e) \in C_1 \setminus B_1$ and $o_1(v_1), o_1(v_2) \in B_1$.*
3. *If s_1 is a conflict part (esp. conflict reason) for rules (r_1, r_2), the set A of all conflict atoms that can be embedded into s_1 is non-empty and covers s_1.*
4. *If s_1 is a conflict reason for rule pair $(r_1, ND(r_2))$, it can be composed of all minimal conflict reasons for $(r_1, ND(r_2))$ that can be embedded into s_1.*
5. *If s_1 is a minimal conflict reason for rule pair $(r_1, ND(r_2))$, its conflict graph S_1 is a subgraph of a deletion component of C_1.*

4   DR/DD Conflict Reasons

Conflict reasons are constructed from conflict part candidates. We distinguish delete-read (dr) from delete-delete (dd) conflict part candidates (and consequently also reasons) by requiring that the dd candidate entails elements that are deleted by both rules, whereas the dr candidate does not.

Definition 6 (dr/dd conflict reason). *Let the rules r_1 and r_2 and a conflict part candidate s_1 for (r_1, r_2) be given.*

1. *s_1 is a dr conflict part candidate for (r_1, r_2) if there exists a morphism $k_{12} : S_1 \to K_2$ such that $le_2 \circ k_{12} = q_{12}$.*
2. *s_1 is a dd conflict part candidate for (r_1, r_2) otherwise.*

A conflict part, atom or (minimal) reason is a dr (dd) conflict part, atom or (minimal) reason, respectively, if it is a dr (dd) conflict part candidate.

A conflict atom consists of either a deleted node or deleted edge with incident preserved nodes (see Fact 1). DR atoms, where the conflict graph consists of a node, might possess incident edges that are deleted not only by the first, but also by the second rule. We say that a dr atom is pure if this is not the case.

Definition 7 (pure dr atom). *Given a conflict reason $s_1 : C_1 \overset{o_1}{\hookleftarrow} S_1 \overset{q_{12}}{\rightrightarrows} L_2$ and a dr atom $s_1' : C_1 \overset{o_1'}{\hookleftarrow} S_1' \overset{q_{12}'}{\rightrightarrows} L_2$ embedded into s_1 via $e : S_1' \to S_1$, then s_1' is pure with respect to s_1 if the conflict graph S_1' consists of an edge, or if S_1' consists of a node x and each edge y in C_1 with source or target node $o_1'(x)$ has a pre-image y' in S_1 with source or target node $e(x)$ s.t. $q_{12}(y') \in le_2(K_2)$.*

In this paper, we consider the general case of rule pairs where both rules may be deleting. This implies that conflicts may arise in both directions. The following definition therefore describes for a given pair of rules and a conflict part candidate how compatible counterparts look like for the reverse direction. It naturally leads to the notion of compatible conflict reasons that may occur in the same conflict in reverse directions. We distinguish compatible counterparts that overlap in at least one deletion item as special case, since they will be important for the dd conflict reason construction.

Definition 8 (compatibility, join, dd overlapping). *Given rules r_1 and r_2 with conflict part candidates s_1 for (r_1, r_2) and $s_2 : C_2 \overset{o_2}{\hookleftarrow} S_2 \overset{q_{21}}{\rightrightarrows} L_1$ for (r_2, r_1) as in Fig. 4.*

1. *Candidates s_1 and s_2 are compatible if the pullbacks $S_1 \overset{a_1}{\hookleftarrow} S' \overset{a_2}{\longrightarrow} S_2$ of $(c_1 \circ o_1, q_{21})$ and $S_1 \overset{a_1'}{\hookleftarrow} S'' \overset{a_2'}{\longrightarrow} S_2$ of $(q_{12}, c_2 \circ o_2)$ are isomorphic via an isomorphism $i : S' \to S''$ such that $a_1' \circ i = a_1$ and $a_2' \circ i = a_2$. We denote a representative of these pullbacks as $s : S_1 \overset{a_1}{\hookleftarrow} S' \overset{a_2}{\hookrightarrow} S_2$.*
2. *Let $S_1 \overset{s_1'}{\hookrightarrow} S \overset{s_2'}{\hookleftarrow} S_2$ be the pushout of s. Morphisms ls_1 and ls_2 are the universal morphisms arising from this pushout and the fact that $c_1 \circ o_1 \circ a_1 = q_{21} \circ a_2$ and $c_2 \circ o_2 \circ a_2 = q_{12} \circ a_1$. Then $L_1 \overset{ls_1}{\hookleftarrow} S \overset{ls_2}{\hookrightarrow} L_2$ as in Fig. 4 is called the join of s_1 and s_2 and S is called the joint conflict graph.*
3. *Compatible conflict part candidates s_1 for (r_1, r_2) and s_2 for (r_2, r_1) are dd overlapping if there do not exist morphisms $k_1 : S' \to K_1$ with $le_1 \circ k_1 = c_1 \circ o_1 \circ a_1$ and $k_2 : S' \to K_2$ with $le_2 \circ k_2 = c_2 \circ o_2 \circ a_2$.*

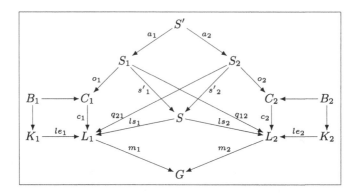

Fig. 4. Compatible conflict part candidates

Conflict parts, atoms, or reasons are (dd overlapping) compatible if the corresponding conflict part candidates are, respectively.

As clarified in the following proposition a dr conflict reason can be responsible on its own for a conflict: if only a dr conflict reason is overlapped by corresponding matches, then we obtain a dr initial conflict. Contrarily, for a dd conflict reason, there exists at least one compatible conflict reason for the reverse rule pair that it can be overlapped with leading to a dd initial conflict. The idea of initial conflicts [10] is that they describe all possible conflicts in a minimal way by overlapping as less elements as possible from both rules.

Proposition 1 (dr/dd conflict reasons and initial conflicts).

- *Given a* dr *conflict reason* $s_1 : C_1 \overset{o_1}{\hookleftarrow} S_1 \overset{q_{12}}{\to} L_2$ *for rule pair* (r_1, r_2), *then the pushout* $(m_1 : L_1 \to K, m_2 : L_2 \to K)$ *of* $L_1 \overset{c_1 \circ o_1}{\hookleftarrow} S_1 \overset{q_{12}}{\to} L_2$ *determines the matches of an dr initial conflict* $(t_1, t_2) = (K \overset{m_1, r_1}{\Longrightarrow} P_1, K \overset{m_2, r_2}{\Longrightarrow} P_2)$ *with the pullback of* $(m_1 \circ c_1, m_2)$ *being isomorphic to* s_1.
- *Given a* dd *conflict reason* s_1 *for rule pair* (r_1, r_2), *then there exists a nonempty set* $DD(s_1)$ *of dd overlapping compatible dd conflict reasons for rule pair* (r_2, r_1) *s.t. for each* s_2 *in* $DD(s_1)$ *the pushout* $(m_1 : L_1 \to K, m_2 : L_2 \to K)$ *of the join of* (s_1, s_2) *determines the matches of an dd initial conflict* $(t_1, t_2) = (K \overset{m_1, r_1}{\Longrightarrow} P_1, K \overset{m_2, r_2}{\Longrightarrow} P_2)$.

Finally, we can conclude from the overapproximation already considered in [2] the following relationship between conflict reasons and overapproximated ones.

Proposition 2 (overapproximating conflict reasons). *If a span* s_1 *is a conflict reason for rule pair* (r_1, r_2), *it is a dr conflict reason for* $(r_1, ND(r_2))$.

5   Characterizing DR/DD Conflict Reasons

Table 2 gives a preview of characterization results for dr/dd conflict reasons described in this section and used for coming up with basic procedures for constructing them in Sect. 6. We start with characterizing dr/dd conflict reasons

Table 2. Characterizing dr/dd conflict reasons for rule pair (r_1, r_2)

Conflict notion	Characterization result
dr conflict reason	covered by pure dr atoms only (Proposition 3)
dd conflict reason	covered by at least one dd or non-pure dr atom and arbitrary number of pure dr atoms (Proposition 3)
dr min. conflict reason	equals min. reason for $(r_1, ND(r_2))$ (Proposition 4)
dd min. conflict reason	composed of min. reasons for $(r_1, ND(r_2))$ being dd conflict part candidates for (r_1, r_2) (Proposition 5)
dr conflict reason	composed of dr min. conflict reasons only (Corollary 1)
dd conflict reason	composed of min. conflict reasons where at least one of which is dd (Corollary 1)

via atoms (Proposition 3). We proceed to characterize dr/dd minimal conflict reasons, showing that we can reuse the constructions for a pair of rules, where the second one is non-deleting (Propositions 4 and 5). We conclude with characterizing dr/dd conflict reasons via minimal ones (Corollary 1). We distinguish dr from dd conflict reasons and learn that the dd case is more involved than the dr case.

Characterizing DR/DD Reasons via Atoms. From the characterization of conflict reasons via atoms (see Fact 1), we can conclude that dr reasons are covered (see Definition 5) by pure dr atoms (see Definition 7). Moreover, each dd reason entails at least one dd atom or non-pure dr atom.

Proposition 3 (dr/dd conflict reason characterization). *A dr conflict reason is covered by pure dr atoms only. On the contrary, a dd conflict reason is covered by at least one dd atom or non-pure dr atom and an arbitrary number of pure dr atoms.*

From the above characterization it follows that it makes sense to distinguish as special case dd conflict reasons that are covered by dd atoms only.

Definition 9 (pure dd conflict reason). *A dd conflict reason is pure if it is covered by dd atoms only.*

Characterizing DR/DD Minimal Reasons. A dr minimal conflict reason for a given rule pair equals the minimal conflict reason for the rule pair, where the second rule of the given rule pair has been made non-deleting.

Proposition 4 (dr minimal conflict reason characterization). *Each dr minimal conflict reason s_1 for rule pair (r_1, r_2) is a dr minimal conflict reason for rule pair $(r_1, ND(r_2))$.*

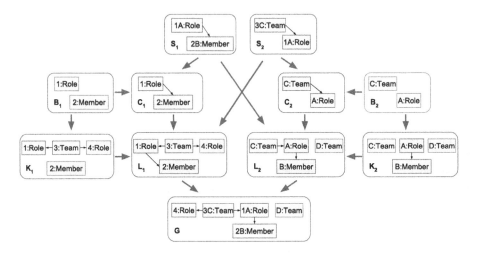

Fig. 5. Two dr minimal conflict reasons for rules *ShiftMember* and *ShiftRole*

We can therefore conclude that the conflict graph of a dr minimal conflict reason is again a subgraph of one deletion component (see Fact 1).

Example 1 (dr minimal conflict reason). Figure 5 shows two dr minimal conflict reasons as examples, one for (r_1, r_2) and one for (r_2, r_1). Note that they do not overlap in elements to be deleted. They are the same minimal reasons as for the cases where the second rule is made non-deleting. For comparison, AGG [11] computes 4 critical pairs for (r_1, r_2) one of which is an initial conflict. Figure 5 shows a critical pair but not the initial conflict. For obtaining the initial conflict, it is enough to overlap merely the dr minimal conflict reason for (r_1, r_2) (see Proposition 1).

A dd minimal conflict reason for a rule pair is composed (Definition 5) of minimal reasons for the rule pair, where the second rule has been made non-deleting.

Proposition 5 (dd minimal conflict reason characterization). *Given a dd minimal conflict reason s_1 for (r_1, r_2), s_1 is composed of a set $M = \{s_i^m \mid i \in I\}$ of minimal conflict reasons for $(r_1, ND(r_2))$. Moreover, each reason in M is a dd conflict part candidate for (r_1, r_2).*

Remember that the conflict graph of each minimal conflict reason for a rule pair, where the second rule is non-deleting, consists of a subgraph of one deletion component. We can therefore conclude that the conflict graph of a dd minimal conflict reason is a subgraph of one or more deletion components.

Example 2 (dd minimal conflict reason). Figure 6 shows an example of a dd minimal conflict reason s_1 for rule pair (*ReplaceM, ReplaceM*). s_1 is a dd conflict reason since S_1 cannot be mapped to K_2 in a suitable way. Furthermore, we see

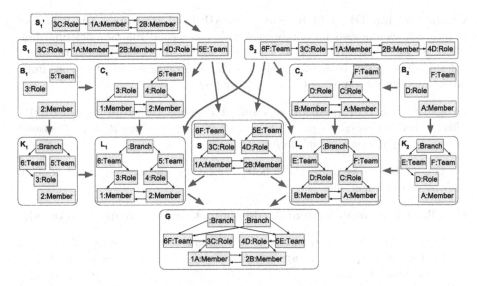

Fig. 6. A dd (minimal) conflict reason for the rule pair (*ReplaceM, ReplaceM*)

that graph G can be constructed such that the completeness condition is fulfilled and m_1 and m_2 are matches. It remains to show that s_1 is indeed minimal. Conflict part candidate s'_1 would also be a promising candidate. The resulting graph G, however, would not merge nodes 4:Role with D:Role. Morphism m_1 would not satisfy the dangling condition then. For a conflict part candidate comprising nodes 2B: Member, 4D: Role, and 5E:Team we can argue similarly. Due to Proposition 5, s_1 has to be composed of minimal conflict reasons for (*ReplaceM*, ND(*ReplaceM*)) which have to be deletion components as shown in [2]. Hence, there are no further possibilities to choose a smaller span than s_1. Note that a dd conflict reason is not always pure. For example, the atom 1A:Member is a (non-pure) dr atom. In addition to this dd minimal conflict reason there exist two more. Their conflict graphs contain the following sets of nodes: {1B:Member, 2A:Member, 3D:Role} and {2A:Member, 4C:Role, 5F:Team}.

For the rule pair (*ReplaceM, ND(ReplaceM)*), four minimal conflict reasons exist instead. Their conflict graphs contain the following sets of nodes: {1B:Member, 2A:Member, 3D:Role}, {2A:Member, 4C:Role, 5F:Team}, {1A:Member, 2B:Member, 3C:Role}, and {2B:Member, 4D:Role, 5E:Team}. While the first two are also conflict graphs of dd minimal conflict reasons for rule pair (*ReplaceM, ReplaceM*), this is not the case for the last two as the dangling condition is not satisfied in those cases. Moreover, the dd minimal conflict reason in Fig. 6 is a conflict reason for (*ReplaceM, ND(ReplaceM)*) but not a minimal one. For comparison, for the rule pair (*ReplaceM, ReplaceM*) AGG [11] computes 71 critical pairs. Figure 6 shows one initial conflict (according to Proposition 1).

Characterizing DR/DD Reasons via Minimal Reasons. The following proposition was proven for a rule pair with the second rule non-deleting, but it can be generalized to the case where the second rule is not necessarily non-deleting. It allows us to construct also for this general case conflict reasons from minimal ones by composing them appropriately.

Proposition 6 (composition of conflict reasons by minimal reasons).
Given a conflict reason s_1 for (r_1, r_2), there is a set of minimal conflict reasons for (r_1, r_2) s_1 is composed of (Definition 5).

This allows to establish the following relationship between dr/dd conflict reasons and minimal ones.

Corollary 1 (composition of dr/dd conflict reasons by minimal ones).

- A dr *conflict reason is composed of dr minimal conflict reasons only.*
- A dd *conflict reason is composed of minimal conflict reasons where at least one of which is dd.*

6 Constructing DR/DD Conflict Reasons

It is known how to construct (minimal) conflict reasons for a rule pair, where the second rule is non-deleting [2]. Proposition 4 tells us that each *dr minimal conflict reason* for a rule pair (r_1, r_2) equals a minimal conflict reason for the rule pair $(r_1, ND(r_2))$. Each such minimal conflict reason for the rule pair $(r_1, ND(r_2))$ that is in addition dr for (r_1, r_2) delivers us a dr minimal conflict reason. From Corollary 1 we know that each *dr conflict reason* is a composition of dr minimal conflict reasons again such that their construction is analogous to the one already presented in [2]. In the following, we construct *dd minimal conflict reasons* from rule pairs, which is much more involved than the dr case.

Definition 10 (composability, composition of conflict part candidates).
Given rules r_1 and r_2 with conflict part candidates s_1 and $s_1' : C_1 \overset{o_1'}{\hookleftarrow} S_1' \overset{q_{21}'}{\rightarrow} L_2$ for (r_1, r_2).

1. *Candidates s_1 and s_1' are composable if the pullbacks $s : S_1 \overset{a_1}{\hookleftarrow} S' \overset{a_2}{\rightarrow} S_1'$ of (o_1, o_1') and $S_1 \overset{a_1'}{\hookleftarrow} S'' \overset{a_2'}{\rightarrow} S_1'$ of (q_{21}, q_{21}') are isomorphic via an isomorphism $i : S' \to S''$ such that $a_1' \circ i = a_1$ and $a_2' \circ i = a_2$. We denote a representative of these pullbacks as s.*
2. *Let $S_1 \overset{s_1'}{\hookrightarrow} S \overset{s_2'}{\hookleftarrow} S_1'$ be the pushout of s. Morphisms ls_1 and ls_2 are the universal morphisms arising from this pushout and the fact that $o_1 \circ a_1 = o_1' \circ a_2$ and $q_{21} \circ a_2 = q_{21}' \circ a_1$. Then $C_1 \overset{ls_1}{\hookleftarrow} S \overset{ls_2}{\hookrightarrow} L_2$ is called the composition of s_1 and s_1'.*

3. *Given a set C of candidates, they are composable for $|C| < 2$. If C is larger, each two of its candidates have to be composable. The composition of all candidates in C is the candidate itself if $|C| = 1$ and the successive composition of its candidates otherwise.*

Conflict parts, atoms, or reasons are composable if the corresponding conflict part candidates are, respectively.

Construction (dd Minimal Conflict Reasons)

Let the rules r_1 and r_2 be given.

- Let CPC_1 be the set of all minimal conflict reasons for $(r_1, ND(r_2))$ which are dd conflict part candidates for (r_1, r_2).
- Given a conflict part candidate s_1, let $CPC_2(s_1)$ be the set of all conflict reasons $s_2 : C_2 \overset{o_2}{\hookleftarrow} S_2 \overset{q_{21}}{\rightarrow} L_1$ for $(r_2, ND(r_1))$ such that s_1 is dd overlapping compatible with s_2.

The set $DDMCR$ of all dd minimal conflict reasons for (r_1, r_2) can be constructed as follows (compare Fig. 4):

1. Let $DDMCR$ be the empty set and $n := 1$.
2. For each subset M of n composable candidates in CPC_1 for which the composition of a subset $M' \subset M$ of $n - 1$ candidates is not in $DDMCR$ yet:
   (a) Compose all candidates in M to a candidate s_1 and construct $CPC_2(s_1)$.
   (b) For each s_2 in $CPC_2(s_1)$:
       - Construct the pushout $L_1 \overset{m_1}{\hookrightarrow} G \overset{m_2}{\hookleftarrow} L_2$ of the join of s_1 and s_2.
       - If m_1 is a match for rule r_1 and m_2 a match for r_2 and if the pullback of $(m_1 \circ o_1, m_2)$ is isomorphic to s_1, then add s_1 to $DDMCR$ and break.
   (c) $n := n + 1$

Remark: Note that a composition of 0 candidates is trivially not in an empty $DDMCR$. This construction terminates since CPC_1 is finite and it has finitely many subsets. n is increased maximally to the size of CPC_1.

Example 3 (Construction of dd min. conflict reason). We construct a dd minimal conflict reason for rule pair (*ReplaceM, ReplaceM*). We start with $n = 1$ and choose s'_1 including conflict graph S'_1 in Fig. 6. It is a min. reason for (*ReplaceM, ND(ReplaceM)*) not belonging to $DDCMR$ yet. As discussed in Example 2, it is not a conflict reason for (*ReplaceM, ReplaceM*). Hence, we cannot find a suitable $s_2 \in CPC_2(s'_1)$. The argumentation for the other minimal conflict reason for (*ReplaceM, ND(ReplaceM)*) is analogous. Hence, we have to set $n = 2$. As s_1 is a composition of two minimal conflict reasons for (*ReplaceM, ND(ReplaceM)*), we choose this candidate next. Figure 6 shows that there is an $s_2 \in CPC_2(s_1)$ such that two matches m_1 and m_2 with the pullback of $(m_1 \circ o_1, m_2)$ being isomorphic to s_1 can be constructed. Hence, s_1 is in $DDMCR$.

Theorem 2 (Correctness dd min. conflict reason construction). *Given two rules r_1 and r_2, the construction above yields dd minimal conflict reasons for (r_1, r_2) only (soundness) and all those (completeness).*

Proof. Soundness: Because of Proposition 5 we know that a dd minimal conflict reason for s_1 for (r_1, r_2) is composed of a set of minimal conflict reasons for $(r_1, ND(r_2))$, where each of them is a dd conflict part candidate for (r_1, r_2). In Step 2 of the construction we select exactly these building bricks for minimal conflict reasons and compose them if composable. We then perform a breadth-first search (w.r.t. size of composition) over all possible compositions of minimal conflict reasons for $(r_1, ND(r_2))$. The search returns all compositions for which we can find a compatible conflict part candidate (see Proposition 2) that leads to a dd reason for (r_1, r_2) (see Proposition 1). We only continue searching for new minimal reasons if we did not find a successful smaller composition already.

Completeness: By checking in the construction for *all* possible combinations of minimal conflict reasons for $(r_1, ND(r_2))$ if we can find a compatible conflict part candidate leading to an initial conflict (see Proposition 1), we find *all* minimal conflict reasons for $(r_1, ND(r_2))$. □

Having constructions for dr/dd minimal conflict reasons at hand, we can compute *dd conflict reasons*. Each dd reason is composed from minimal reasons, where at least one of them is dd (see Corollary 1). Their construction is thus analogous to the one for dd minimal conflict reasons with the following two differences: (1) Instead of CPC_1 we have the set MCR_1 of minimal dr/dd reasons for (r_1, r_2). (2) Step 2 considers each set of n composable minimal reasons in MCR_1 with at least one of them dd, no matter if the composition of a subset is already present in the result set or not (since we do not need minimality). Soundness and completeness follows analogous to the proof of Theorem 2 based on Corollary 1 instead of Proposition 5 and omitting the argument for minimality.

7   Related Work and Conclusion

Our paper continues a recent line of research on conflict and dependency analysis (CDA) for graph transformations aiming to improve on the previous CDA technique of critical pair analysis (CPA). Originally inspired by the CPA in term and term graph rewriting [3], the CPA theory has been extended to graph transformation and generally, to $\mathcal{M}$-adhesive transformation systems [1,12].

Azzi et al. [13] conducted similar research to identify root causes of conflicting transformations as initiated in [9] and continued in [5]. Their work is based upon an alternative characterization of parallel independence [14] that led to a new categorical construction of initial transformation pairs. The most important difference is that we define our conflict notions (including the dr/dd characterization) for rule pairs instead of transformation pairs [13] with the aim of coming up with efficient CDA. Moreover, we consider conflict atoms and (minimal) reasons, whereas Azzi et al. [13] focus conflict reasons (in our terminology).

In this paper, we extend the foundations for computing conflicts and dependencies for graph transformations in a multi-granular way. In particular, our earlier work relied on an over-approximation of (minimal) conflict reasons; we assumed a non-deleting version of the second rule of the considered rule pair as input. In contrast, our present work introduces a new constructive characterization of (minimal) conflict reasons distinguishing dr from dd reasons and we present a basic computation procedure that is sound and complete. Building on our recent work [2], we now support precise conflict computation for any given granularity level, from binary (conflict atom) over medium (minimal conflict reason) to fine (conflict reason). Future work is needed to implement the presented constructions, to evaluate efficiency and to investigate usability.

Acknowledgements. This work was partially funded by the German Research Foundation, project "Triple Graph Grammars (TGG) 2.0: Reliable and Scalable Model Integration".

References

1. Ehrig, H., Ehrig, K., Prange, U., Taentzer, G.: Fundamentals of Algebraic Graph Transformation. Monographs in Theoretical Computer Science. An EATCS Series. MTCSAES. Springer, Heidelberg (2006). https://doi.org/10.1007/3-540-31188-2
2. Lambers, L., Strüber, D., Taentzer, G., Born, K., Huebert, J.: Multi-granular conflict and dependency analysis in software engineering based on graph transformation. In: International Conference on Software Engineering (ICSE). pp. 716–727. ACM (2018). Extended version. www.uni-marburg.de/fb12/swt/forschung/publikationen/2018/LSTBH18-TR.pdf
3. Plump, D.: Critical pairs in term graph rewriting. In: Prívara, I., Rovan, B., Ruzička, P. (eds.) MFCS 1994. LNCS, vol. 841, pp. 556–566. Springer, Heidelberg (1994). https://doi.org/10.1007/3-540-58338-6_102
4. Lambers, L., Born, K., Kosiol, J., Strüber, D., Taentzer, G.: Granularity of conflicts and dependencies in graph transformation systems: a two-dimensional approach. J. Log. Algebr. Methods Program. **103**, 105–129 (2019)
5. Born, K., Lambers, L., Strüber, D., Taentzer, G.: Granularity of conflicts and dependencies in graph transformation systems. In: de Lara, J., Plump, D. (eds.) ICGT 2017. LNCS, vol. 10373, pp. 125–141. Springer, Cham (2017). https://doi.org/10.1007/978-3-319-61470-0_8
6. Lambers, L., Kosiol, J., Strüber, D., Taentzer, G.: Exploring conflict reasons for graph transformation systems: Extended version (2019). https://www.uni-marburg.de/fb12/arbeitsgruppen/swt/forschung/publikationen/2019/LKST19-TR.pdf
7. Beck, K., et al.: Manifesto for Agile software development (2001)
8. Arendt, T., Biermann, E., Jurack, S., Krause, C., Taentzer, G.: Henshin: advanced concepts and tools for in-place EMF model transformations. In: Petriu, D.C., Rouquette, N., Haugen, Ø. (eds.) MODELS 2010. LNCS, vol. 6394, pp. 121–135. Springer, Heidelberg (2010). https://doi.org/10.1007/978-3-642-16145-2_9
9. Lambers, L., Ehrig, H., Orejas, F.: Efficient conflict detection in graph transformation systems by essential critical pairs. Electr. Notes Theor. Comput. Sci. **211**, 17–26 (2008)

10. Lambers, L., Born, K., Orejas, F., Strüber, D., Taentzer, G.: Initial conflicts and dependencies: critical pairs revisited. In: Heckel, R., Taentzer, G. (eds.) Graph Transformation, Specifications, and Nets. LNCS, vol. 10800, pp. 105–123. Springer, Cham (2018). https://doi.org/10.1007/978-3-319-75396-6_6
11. AGG: Attributed Graph Grammar system. http://user.cs.tu-berlin.de/~gragra/agg/
12. Ehrig, H., Padberg, J., Prange, U., Habel, A.: Adhesive high-level replacement systems: a new categorical framework for graph transformation. Fundam. Inform. **74**(1), 1–29 (2006)
13. Azzi, G.G., Corradini, A., Ribeiro, L.: On the essence and initiality of conflicts. In: Lambers, L., Weber, J. (eds.) ICGT 2018. LNCS, vol. 10887, pp. 99–117. Springer, Cham (2018). https://doi.org/10.1007/978-3-319-92991-0_7
14. Corradini, A., et al.: On the essence of parallel independence for the double-pushout and sesqui-pushout approaches. In: Heckel, R., Taentzer, G. (eds.) Graph Transformation, Specifications, and Nets. LNCS, vol. 10800, pp. 1–18. Springer, Cham (2018). https://doi.org/10.1007/978-3-319-75396-6_1

Unfolding Graph Grammars
with Negative Application Conditions

Andrea Corradini[1(✉)], Maryam Ghaffari Saadat[2], and Reiko Heckel[2]

[1] Dipartimento di Informatica, University of Pisa, Pisa, Italy
andrea@di.unipi.it
[2] Department of Informatics, University of Leicester, Leicester, UK
{mgs17,rh122}@leicester.ac.uk

Abstract. The unfolding of a graph grammar provides a compact and comprehensive representation of its behaviour, serving both as a semantic model and as the basis for scalable analysis techniques. We study the extension of the theory of unfolding to grammars with negative application conditions (NACs), discuss the challenges with the general case of NACs consisting of complex graph patterns and how they could be avoided by restricting to simpler, incremental NACs.

Keywords: Graph grammars · Unfolding semantics ·
Negative application conditions · Incremental NACs

1 Introduction

Graph grammars provide a natural way of modelling computations over graphs, in order to formalise semantics, support analysis, or solve combinatorial problems. Arguably the most comprehensive representation of the concurrent and non-deterministic behaviour a graph grammar is its unfolding [1–3]. This can be computed incrementally and represents in one structure the branching computations of the grammar in what could be described as a partial-order variant of its derivation tree.

Negative application conditions (NACs) [9] increase the expressiveness of rules by allowing to specify preconditions requiring the absence of nodes, edges or patterns in the context of the match. While not formally increasing the computational power of double-pushout (DPO) graph grammars [5,6] (which are already Turing-complete) they support a more high-level way of modelling. By avoiding the need to encode negative information into the graph structure they tend to reduce the size of models and thus make analysis more feasible.

In the presence of NACs, the concurrent and non-deterministic behaviour of a graph grammar becomes significantly more complex. While, traditionally, dependencies and conflicts between transformations are based on how their left- and right-hand sides overlap in given or derived graphs, and if these overlaps include elements created or deleted by one rule and required or preserved by

© Springer Nature Switzerland AG 2019
E. Guerra and F. Orejas (Eds.): ICGT 2019, LNCS 11629, pp. 93–110, 2019.
https://doi.org/10.1007/978-3-030-23611-3_6

the other, NACs entail new types of conflicts and dependencies. In particular, a *create-forbid* conflict exists if one rule creates part of a structure that is forbidden by the NAC of another, thus disabling it. A *delete-enable* dependency is caused by one rule enabling another one by deleting elements that previously caused a violation of the second rule's NAC.

The added complexity here is due to the fact that, even in a given derivation, there may be several possible transformations contributing to the creation or destruction of a substructure matching a negative condition. That means, a delete-enable dependency may exist not with a unique transformation but with a set of alternative transformations each deleting part of that forbidden structure. Dually, a create-forbid conflict can arise with any of a number of transformations creating part of such a structure. That means, intuitively, these relations have a disjunctive flavour in contrast to the conjunctive nature of the classical causal dependency where all transformations establishing part of a rule's match have to be present for this rule to be applied. In addition, a rule with a NAC may be applied to a given match either before the NAC occurs in the derivation or after it has been destroyed, adding another level of disjunction.

We address the problem of unfolding conditional graph grammars generalising the approach by [14] for the unfolding of Place/Transition Petri nets with read and inhibitor arcs. The idea here is to resolve the disjunction mentioned above by introducing so-called assignments that fix, for each NAC occurrence, a transformation creating or destroying part of this occurrence. The causal history of a rule can then be considered relative to such an assignment.

In analogy to [14] we define a notion of conditional occurrence graph grammar assuming the existence of a suitable assignment for every rule and show how to create such an occurrence grammar by unfolding a given conditional graph grammar. The unfolding is equipped with a morphism to the original grammar, and as a consequence each derivation of the unfolding is mapped to a derivation of the original grammar. It remains beyond the scope this paper, but a topic of future work, to prove that the resulting construction gives rise to a co-reflection. This means that not only the unfolding preserves the behaviour, but it is also maximal with this property, in the sense that each morphism from an occurrence grammar to the original grammar factorizes uniquely through the unfolding.

Somewhat surprisingly, our approach works for both the incremental case (the more direct generalisation of a net with read and inhibitor arcs, because the inhibitor arc only points to a single place) and the general case of complex forbidden patterns. However, this makes the choice of assignments in the construction of the unfolding highly non-deterministic calling into question its use as an efficient computational structure. We discuss how a restriction to incremental NACs leads to a canonical choice of assignments bringing the complexity of the unfolding construction closer to the unconditional case.

Related Work. So far, only some of the more basic elements of the theory of concurrency of graph grammars have been lifted to the case with NACs, including the classical Local Church Rosser and Concurrency theorems [9,10] and the analysis of critical pairs [11]. Graph processes, a deterministic form of unfolding, have been studied for grammars with NACs in [12]. In [13] two of the authors established the existence of canonical derivations for systems with *incremental NACs* which, in the case of graphs, are limited to forbidding individual nodes, or edges attached to existing nodes (and their opposite source or target node). The theory of approximate unfolding [4] in the DPO approach has been used to implement scalable verification [7] and optimisation [8].

The paper is organized as follows: in the following section we introduce the main concepts of conditional graph grammars; in Sect. 3, we review the notion of occurrence grammar and of unfolding in the unconditional case. Section 4 introduces conditional occurrence grammars and Sect. 5 presents the corresponding unfolding construction. Section 6 is devoted to a discussion of the case with incremental NACs and Sect. 7 concludes the paper.

2   Basic Definitions

This section summarizes the basic definitions of typed graph grammars [15] based on the DPO approach [5,6] extended by negative application conditions (NACs) [9]. Formally, a *(directed, unlabelled) graph* is a tuple $G = \langle N, E, s, t \rangle$, where N is a set of *nodes*, E is a set of *arcs*, and $s, t : E \to N$ are the *source* and *target* functions. A *graph morphism* $f : G \to G'$ is a pair of functions $f = \langle f_N : N \to N', f_E : E \to E' \rangle$ preserving sources and targets, i.e., such that $f_N \circ s = s' \circ f_E$ and $f_N \circ t = t' \circ f_E$.

The category of graphs and graph morphisms is denoted by **Graph**. Given a graph TG, called *type graph*, a *TG-typed (instance) graph* is a pair $\langle G, t_G \rangle$, where G is a graph and $t_G : G \to TG$ is a morphism. A morphism between typed graphs $f : \langle G_1, t_{G_1} \rangle \to \langle G_2, t_{G_2} \rangle$ is a graph morphisms $f : G_1 \to G_2$ consistent with the typing, i.e., such that $t_{G_1} = t_{G_2} \circ f$. The category of TG-typed graphs and typed graph morphisms is denoted by **TG-Graph**.

A *rule* $p = (L \xleftarrow{l} K \xrightarrow{r} R)$ consists of a span of two injective graph morphisms l and r. A rule is *consuming* if l is not an isomorphism: all along the paper rules are assumed to be consuming. A *redex* in a graph G is a pair $\langle p, m \rangle$, where p is a rule and $m : L \to G$ is an injective graph morphisms, called a *match*.

Given a redex $\langle p, m \rangle$, a *double-pushout (DPO) transformation* $G \xRightarrow{p,m} H$ from G to H exists if we can construct a diagram such as (1) where both squares are pushouts in **TG-Graph**.

$$\begin{array}{ccccc} L & \xleftarrow{\ l\ } & K & \xrightarrow{\ r\ } & R \\ \scriptstyle m\big\downarrow & & \scriptstyle d\big\downarrow & & \big\downarrow\scriptstyle m^* \\ G & \xleftarrow{\ g\ } & D & \xrightarrow{\ h\ } & H \end{array} \qquad (1)$$

The applicability of rules can be restricted by negative application conditions. For a rule p as above, a *(negative) constraint* over p is a monomorphism $n : L \to N$. A match $m : L \to G$ satisfies n if there is no monomorphism $q : N \to G$ such that $q \circ n = m$. A negative application condition (NAC) over a rule p is a set of constraints over p. A match $m : L \to G$ *satisfies* a NAC ϕ over p if m satisfies every constraint in ϕ. A *graph grammar* $\mathcal{G} = \langle TG, G_{in}, P, \pi \rangle$ consists of a type graph TG, a TG-typed input graph G_{in}, a set of rule names P and a function π assigning to each $p \in P$ a rule $\pi(p) = (L_p \xleftarrow{l} K_p \xrightarrow{r} R_p)$.

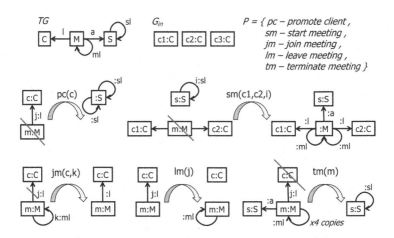

Fig. 1. Conditional graph grammar

A *conditional graph grammar* $CG = \langle TG, G_{in}, P, \pi, \Phi \rangle$ adds a function Φ providing for each $p \in P$ a NAC $\Phi(p)$ over $\pi(p)$. We denote by Φ_{CG} the total set of constraints of the grammar, i.e., $\Phi_{CG} = \{\langle q, n \rangle \mid q \in P, n \in \Phi(q)\}$. A transformation with NACs is called *conditional transformation*.

A *derivation* in G is a finite sequence of transformations $s = (G_{in} = G_0 \xrightarrow{p_1, m_1} \cdots \xrightarrow{p_n, m_n} G_n)$ with $p_i \in P$.[1] A *conditional derivation* in CG is a derivation in G such that each $G_{i-1} \xrightarrow{p_i, m_i} G_i$ is a conditional transformation, that is, its match m_i satisfies $\Phi(p_i)$.

Example 1 (Conditional Graph Grammar). Figure 1 shows an example of a conditional graph grammar CG modelling an online meeting system where Clients can be promoted to act as Servers, who can then start Meetings between two Clients. Up to two additional Clients can join each meeting, and Clients can leave the Meeting independently, for an empty meeting to be terminated. The number for Meetings per Server and Clients per Meeting are limited by loops on Server and Meeting nodes, respectively. They also ensure that the system is consuming, i.e., each rule deletes at least one element of its left-hand side. When terminating, the meeting rule $tm(m)$ deletes the four loops attached to m when no Client is connected.

We use a condensed notation for rules, omitting the interface graph of the span (which can be reconstructed as the intersection of the left- and right-hand sides) and integrating the negative elements into the left-hand side. For example, in rule $pc(c)$ the left hand side $L_{pc(c)}$ consists of a single Client node while its negative constraint $\hat{L}_{pc(c)}$ consists of the Client node, the Meeting node and the edge between them, indicating that the rule can only be applied if Client c is not in a Meeting. The negative part of the constraint is $\hat{L}_{pc(c)} \setminus L_{pc(c)}$, given by the Meeting node and the edge only.

---

[1] For the sake of simplicity we often leave the application of function π implicit.

It is well known [9] that NACs can express the dangling condition of the DPO approach. The idea is to add to every rule p a constraint for each deleted node v, requesting that v is not the source or target of edges beside those in p's left-hand side. If this is true for all rules of a conditional grammar CG, we say that its NACs *subsume the dangling condition*. Examples can be seen in rules $pc(c)$ and $tm(c, m)$ above. In the rest of the paper we assume this of all conditional grammars.

3   Occurrence Grammars

As a prerequisite to considering conditional grammars, in this section we recall the notion of non-deterministic occurrence grammar and describe informally the unfolding construction [1].

A grammar $\mathcal{G} = \langle TG, G_{in}, P, \pi \rangle$ is *(strongly) safe* if, for all H such that $G_{in} \Rightarrow^* H$, H has an injective typing morphism. The basic observation is that typed graphs having an injective typing morphism can be safely identified with the corresponding subgraphs of the type graph (just thinking of injective morphisms as inclusions). Therefore, in particular, each graph $\langle G, t_G \rangle$ reachable in a safe grammar can be identified with the subgraph $t_G(G)$ of the type graph TG. For a safe grammar $\mathcal{G} = \langle TG, G_{in}, P, \pi \rangle$, the set of its *elements* is defined as $Elem(\mathcal{G}) = TG_E \cup TG_N \cup P$, assuming without loss of generality that the three sets are mutually disjoint.

Using a net-like language, we speak of *pre-set* $^\bullet q$, *context* $\underline{q}$ and *post-set* $q^\bullet$ of a rule q, defined as the disjoint sets of elements deleted, preserved and created by the rule. Similarly for a node or arc x in TG we write $^\bullet x$, $\underline{x}$ and $x^\bullet$ to denote the disjoint sets of rules which produce, preserve and consume x.[2]

Definition 1 (causal and asymmetric conflict relations). *The* causal rela-tion *of a safe grammar $\mathcal{G}$ is the binary relation $<$ over $Elem(\mathcal{G})$ defined as the least transitive relation satisfying: for any node or arc x in the type graph TG, and for rules $q_1, q_2 \in P$*

1. *if $x \in {}^\bullet q_1$ then $x < q_1$;*
2. *if $x \in q_1{}^\bullet$ then $q_1 < x$;*
3. *if $q_1{}^\bullet \cap \underline{q_2} \neq \emptyset$ then $q_1 < q_2$;*

As usual $\leq$ is the reflexive closure of $<$. Moreover, for $x \in Elem(\mathcal{G})$ we denote by $\lfloor x \rfloor$ the set of causes of x in P, namely $\{q \in P : q \leq x\}$.

The asymmetric conflict relation *of a grammar $\mathcal{G}$ is the binary relation $\nearrow$ over the set of rules, defined by:*

1. *if $\underline{q_1} \cap {}^\bullet q_2 \neq \emptyset$ then $q_1 \nearrow q_2$;*
2. *if $^\bullet q_1 \cap {}^\bullet q_2 \neq \emptyset$ and $q_1 \neq q_2$ then $q_1 \nearrow q_2$;*
3. *if $q_1 < q_2$ then $q_1 \nearrow q_2$.*

---

[2] For the sake of conciseness, we depart slightly from [1] by providing the next defini-tion for safe grammars only.

Given two productions q_1 and q_2 of a safe grammar, if $q_1 < q_2$ then in any derivation where q_2 occurs, q_1 must occur before q_2, as it produces some item necessary for q_2. Instead $q_1 \nearrow q_2$ states that in any derivation where *both* occur, q_1 must precede q_2, but q_2 could occur alone. This is the case when (first clause) q_2 deletes an item that is preserved by q_1.

A *nondeterministic occurrence grammar* is a safe grammar where the causal relation is acyclic, and which represents, in a branching structure, several possible computations starting from its initial graph and using each rule at most once.

Definition 2 ((nondeterministic) occurrence grammar [1]). *A* (nondeterministic) occurrence grammar *is a safe grammar* $\mathcal{O} = \langle TG, G_{in}, P, \pi \rangle$ *where*

1. *for each rule* $q : \langle L, t_L \rangle \xleftarrow{l} \langle K, t_K \rangle \xrightarrow{r} \langle R, t_R \rangle$, *the typing morphisms* t_L, t_K *and* t_R *are injective.*
2. *its causal relation* $\leq$ *is a partial order, and for any* $q \in P$, *the set* $\lfloor q \rfloor$ *is finite and asymmetric conflict* $\nearrow$ *is acyclic on* $\lfloor q \rfloor$;
3. *the initial graph* G_{in} *coincides with the set* $Min(\mathcal{O})$ *of minimal elements of* $\langle Elem(\mathcal{O}), \leq \rangle^3$ *(with the graphical structure inherited from* TG *and typed by the inclusion);*
4. *each arc or node* x *in* TG *is created by at most one rule in* P: $|{}^\bullet x| \leq 1$.

Since the initial graph of an occurrence grammar $\mathcal{O}$ *is determined by* $Min(\mathcal{O})$, *we often do not mention it explicitly.*

Example 2 (Occurrence Grammar). Figure 2 shows an occurrence grammar $\mathcal{OG}$ based on the grammar $\mathcal{G}$ underlying the conditional grammar $\mathcal{CG}$ of Example 1. That means, in $\mathcal{G}$ we keep type graph and rules of $\mathcal{OG}$ but drop all application conditions. The type graph at the bottom of Fig. 2 contains all graph elements required for the application of a selection of rules of $\mathcal{G}$, represented by the rule occurrences listed above it. These occurrences are the rule names of $\mathcal{OG}$, shown by means of their names in $\mathcal{G}$, using parameters to indicate their matches. For example, $sm(c1, c3, 2_1)$ represents an application of rule $sm(c1, c2, 1)$ to Server $s2$ and Clients $c1, c3$, consuming the loop 2_1 and producing a meeting $m21$ with links 21-1, 21-3 to $c1, c3$, respectively. That means, pre-set, context and post-set are given by ${}^\bullet sm(c1, c3, 2_1) = \{2_1\}$, $sm(c1, c3, 2_1) = \{c1, c3, s2\}$ and $sm(c1, c3, 2_1)^\bullet = \{m21, 21\text{-}1, 21\text{-}3, 21_1, 21_2, 1\}$. Independently, $sm(c1, c3, 2_2)$ produces a meeting $m22$ linking the same clients via 22-1, 22-3 after consuming loop 2_2. In particular, its pre-set and context are given by ${}^\bullet sm(c1, c3, 2_2) = \{2_2\}$, $sm(c1, c3, 2_2) = \{c1, c3, s2\}$, so according to Definition 1 they are neither related by $<$ nor by $\nearrow$.

Instead, with pre-set, context and post-set for $pc(c2)$ given by ${}^\bullet pc(c2) = \{c2\}$, $pc(c2) = \emptyset$ and $pc(c2)^\bullet = \{s2, 2_1, 2_2\}$, both $pc(c2) < sm(c1, c3, 2_1)$ and $pc(c2) < sm(c1, c3, 2_2)$ because by clause 2 of the definition of $<$, $pc(c2) < 2_i$ and by clause 1, $2_i < sm(c1, c3, 2_i)$. Similarly, $jm(c1, 2i_1), jm(c1, 2i_2)$ are independent, but $sm(c1, c3, 2_i) < jm(c1, 2i_1), jm(c1, 2i_2)$. Furthermore, we have

---

^{3} Notice that $Min(\mathcal{O}) \subseteq N_{TG} \cup E_{TG}$, i.e., it does not contain rules, since the grammar is consuming.

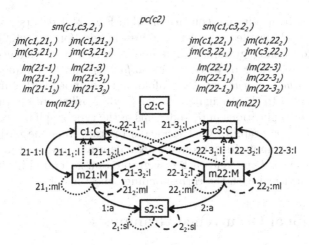

Fig. 2. Occurrence grammar $\mathcal{OG}$. For readability the type graph does not show loops on $m2i$ created by $lm(j)$ occurrences nor loops on $s2$ created by $tm(m)$ occurrences.

mutual conflicts $\nearrow$ among all occurrences $jm(cl, 2i_1)$ consuming $2i_1$ and all occurrences $jm(cl, 2i_2)$ consuming $2i_2$. All occurrences of $lm(j)$ are independent of each other because they each consume their own link. Occurrences $tm(m2i)$ consume $m2i$, i.e., $m2i \in {}^\bullet tm(m2i)$, and since $m2i \in jm(cl, 2i_k)$, we have $jm(cl, 2i_k) \nearrow tm(m2i)$ by clause 1 of $\nearrow$.

Definition 3 (concurrent graph). *A subgraph G of the type graph TG of an occurrence grammar is* concurrent *if*

1. $\neg(x < y)$ *for all* $x, y \in G$
2. $\lfloor G \rfloor$ *is finite*
3. $\nearrow$ *is acyclic on* $\lfloor G \rfloor$
4. *for all* $e \in TG$ *and* $n \in \{s(e), t(e)\}$, *if* $n^\bullet \cap \lfloor G \rfloor \neq \emptyset$ *and* ${}^\bullet e \subseteq \lfloor G \rfloor$ *then* $e^\bullet \cap \lfloor G \rfloor \neq \emptyset$

Condition 4. Ensures the satisfaction of the dangling condition for all rules in $\lfloor G \rfloor$. Concurrent graphs represent the coverable graphs of the grammar [1].

Proposition 1 (concurrent graph). *A graph $G \subseteq TG$ is concurrent if and only if it is* coverable *in $\mathcal{O}$, i.e. there is derivation $Min(\mathcal{O}) \Rightarrow^* H$ with $G \subseteq H$.*

The idea is that, for every concurrent graph G one can find a derivation $Min(\mathcal{O}) \Rightarrow^* H$ which applies exactly once every rule in $\lfloor G \rfloor$, in any order consistent with $(\nearrow_{\lfloor G \rfloor})^*$. Vice versa for each derivation $Min(\mathcal{O}) \Rightarrow^* G$ in $\mathcal{O}$, the set of rules it applies contains $\lfloor G \rfloor$ and their order is compatible with $\nearrow^*$. Therefore reachable graphs are concurrent. Furthermore, each subgraph of a concurrent graph is concurrent as well, thus so are all coverable graphs.

For example in the type graph of Fig. 2, each subgraph consisting of an l-edge and its Meeting and Client node is concurrent and thus contained in a graph reachable by a suitable sequence of rules. E.g., 21-1 results from applying $pc(c2); sm(c1, c3, 2_1)$ while each 12-1$_i$ requires, in addition, $jm(c1, 21_i)$.

Given a consuming grammar $\mathcal{G}$, a nondeterministic occurrence grammar describing the behaviour of $\mathcal{G}$, called its *unfolding*, can be obtained with the so-called *unfolding construction*. The unfolding is equipped with a morphism of graph grammars u to the original grammar $\mathcal{G}$ which allows to see rules in $\mathcal{O}$ as instances of rule applications in $\mathcal{G}$, and items of the type graph of $\mathcal{O}$ as instances of items of the type graph of $\mathcal{G}$. The idea is to start from the initial graph of the grammar, apply its rules in all possible ways, recording in the unfolding each occurrence of rule and each new graph item generated in the rewriting process, both enriched with the corresponding causal history.

4   Conditional Occurrence Grammars

We define morphisms of conditional graph grammars such that they preserve behaviour. For a graph morphism $f_{TG} : TG_1 \rightarrow TG_2$, we denote by $f_{TG}^>$ the covariant retyping functor from TG_1-typed graphs to TG_2-typed graphs, defined as $f_{TG}^>(\langle G_1, t_{G_1}\rangle) = \langle G_1, f_{TG} \circ t_{G_1}\rangle$.

Definition 4 (CGG morphisms). *Given conditional graph grammars $\mathcal{CG}_i = \langle TG_i, G_i, P_i, \pi_i, \Phi_i \rangle$ for $i = 1, 2$ a CGG morphism $f = \langle f_{TG}, f_P \rangle : \mathcal{CG}_1 \rightarrow \mathcal{CG}_2$ consists of a graph morphism $f_{TG} : TG_1 \rightarrow TG_2$ and a mapping $f_P : P_1 \rightarrow P_2$ such that*

- *$G_2 = f_{TG}^>(G_1)$*
- *$\pi_2(f_P(p)) = f_{TG}^>(\pi_1(p))$ for all $p \in P_1$*
- *$f_{TG}^>(L_p \rightarrow N) \in \Phi_2(f_P(p)) \Rightarrow (L_p \rightarrow N) \in \Phi_1(p)$.*

The category of conditional graph grammars and CGG morphisms is **CGG**.

Proposition 2 (CGG morphisms preserve derivations). *Morphisms of conditional graph grammars preserve derivations.*

Proof. We show that given a transformation $G \xrightarrow{p,m} H$ in $\mathcal{CG}_1$ we obtain a transformation $f_{TG}^>(G) \xrightarrow{f_P(p), f_{TG}^>(m)} f_{TG}^>(H)$ in $\mathcal{CG}_2$. Then, this mapping extends to derivations of length >1 by concatenation.

Functor $f_{TG}^>$ preserves pushouts. That means, for the underlying unconditional grammars, applying $f_{TG}^>$ to the DPO diagram of a transformations $G \xrightarrow{p,m} H$ in $\mathcal{G}_1$ we obtain the DPO diagram of a transformation $f_{TG}^>(G) \xrightarrow{f_P(p), f_{TG}^>(m)} f_{TG}^>(H)$ in $\mathcal{G}_2$.

Assume that m satisfies $\Phi_1(p)$, i.e., $G \xrightarrow{p,m} H$ is a transformation in $\mathcal{CG}_1$. We proceed by contradiction, assuming that there is a constraint $n_2 : f_{TG}^>(L_p) \rightarrow N_2 \in \Phi_2(f_P(p))$ not satisfied by $f_{TG}^>(m)$, i.e., there exists an injective morphism $q : N_2 \rightarrow f_{TG}^>(G)$ such that $q \circ n_2 = f_{TG}^>(m)$. In this case, there exists a morphism $n_1 : L_p \rightarrow N_1$, typed over TG_1, such that $f_{TG}^>(n_1) = n_2$. In particular, TG_1-typed graph N_1 is the same as N_2 with typing $t_{N_1} = t_G \circ q$ (*). This composition is well-defined because G and $f_{TG}^{>}(G)$ are the same graph with different typing morphisms, i.e., $t_{f_{TG}^{>}(G)} = f_{TG} \circ t_G$. TG_2-typed morphism n_2 is also typed over TG_1 because $t_{N_1} \circ n_2 = t_G \circ q \circ n_2 = t_{L_p}$ based on the analogous observation that L_p and $f_{TG}^{>}(L_p)$ are the same graph, but for typing. We let n_1 be this TG_1-typed version of n_2, so $f_{TG}^{>}(n) = n_2$.

By the third clause in Definition 4, $n_1 \in \Phi_1(p)$. Now, if $f_{TG}^{>}(m)$ does not satisfy n_2, there exists an injective morphism $q : f_{TG}^{>}(N_1) \to f_{TG}^{>}(G)$ such that $q \circ f_{TG}^{>}(n_1) = f_{TG}^{>}(m)$. Morphism q is also a TG_1-typed $q : N_1 \to G$ if $t_G \circ q = t_{N_1}$. However, this is just as t_{N_1} was defined in $(*)$ above. Hence m does not satisfy $n_1 \in \Phi_1(p)$, contradicting the assumption. □

Following [14], we introduce the notion of *assignment* that will be pivotal in the definition of conditional occurrence grammar. An assigment determines for each constraint $\langle q, \hat{L}_q \rangle$[4] of a rule q, which rule guarantees that the constraint is satisfied when q is applied.

Definition 5 (assignment). *Let CG be a safe conditional grammar. An assignment for CG is a function $\rho : \Phi_{CG} \to P$ such that, for all $\langle q, \hat{L}_q \rangle \in \Phi_{CG}$, $\rho(\langle q, \hat{L}_q \rangle) \in {}^{\bullet}(\hat{L}_q \setminus L_q) \cup (\hat{L}_q \setminus L_q)^{\bullet}$.*

Given a constraint $n = \langle q, \hat{L}_q \rangle$, an assignment ρ associates with n either a rule that deletes one or more items forbidden by n, guaranteeing that q can be applied after $\rho(n)$, or a rule that creates one or more items forbidden by n, meaning that q can be applied before $\rho(n)$. Therefore once an assignment ρ for CG is fixed, new dependencies among their rules arise, that we denote by $<_\rho^n$ and $\nearrow_\rho^n$. More precisely, we define:

- $\rho(\langle q, \hat{L}_q \rangle) <_\rho^n q$ if $\rho(\langle q, \hat{L}_q \rangle) \in (\hat{L}_q \setminus L_q)^{\bullet}$, and
- $q \nearrow_\rho^n \rho(\langle q, \hat{L}_q \rangle)$ if $\rho(\langle q, \hat{L}_q \rangle) \in {}^{\bullet}(\hat{L}_q \setminus L_q)$ and $\rho(\langle q, \hat{L}_q \rangle) \neq q$.

The notation suggests in the first case a proper causality (if q is applied, then $\rho(\langle q, \hat{L}_q \rangle)$ should have been applied before), while in the second case we have an asymmetric conflict (if both q and $\rho(\langle q, \hat{L}_q \rangle)$ are applied, q must be applied before). The second clause also considers the *self-disabling* case where q itself generates (part of) its constraint, in which case no new dependencies arise.

We define extended relations $<_\rho$ and $\nearrow_\rho$, integrating conflicts and dependencies from negative constraints, as $<_\rho = (< \cup <_\rho^n)^+$ and $\nearrow_\rho = <_\rho \cup \nearrow \cup \nearrow_\rho^n$. For $x \in Elem(CG)$ we denote by $\lfloor x \rfloor_\rho$ the set $\{q \in P \mid q \leq_\rho x\}$ and by $\lfloor X \rfloor_\rho$ the extension to sets $X \subseteq Elem(CG)$.

A conditional grammar CG is *safe* if its underlying grammar G is, and all its constraints have injective typing morphisms. As before, we assume inclusions and represent a constraint $n : L \to \hat{L}$ by its target graph $\hat{L}$ only.

Definition 6 (conditional occurrence grammar). *A conditional occurrence grammar is a safe grammar $CO = \langle TG, G_{in}, P, \pi, \Phi \rangle$ where*

---

[4] We denote a constraint $\langle q, L_q \xrightarrow{n} \hat{L}_q \rangle$ simply as $\langle q, \hat{L}_q \rangle$ assuming that n is an inclusion.

1. *for each rule q, its typing morphisms t_{L_q}, t_{K_q} and t_{R_q} are injective, and for each constraint $\langle q, \hat{L}_q \rangle \in \Phi_{\mathcal{CG}}$ graph $\hat{L}_q$ is typed injectively*
2. *for all $q \in P$ there exists an assignment ρ such that $(\nearrow_\rho)_{\lfloor q \rfloor_\rho}$ is acyclic and $\lfloor q \rfloor_\rho$ is finite*
3. *the (unconditional) grammar $\mathcal{O} = \langle TG, G_{in}, P, \pi \rangle$ satisfies the conditions of Definition 2 (3) and (4).*

The category of conditional occurrence grammars and CGG morphisms is denoted by **COG**.

Example 3 (Conditional Occurrence Grammar). We extend Example 2 by constraints associated with rule occurrences. For example, none of the 12 l-edges are constraints of $pc(c2)$ because they are not attached to $c2$. Instead, any of the three edges $j : l$ from $m21$ to $c1$ is a constraint $\hat{L}(j)$ of $jm(c1, 21_1)$.

An assignment ρ discharges existing constraints, justifying the applicability of rules for new dependencies or conflicts. Assignment $\rho(\langle jm(c1, 21_1), \hat{L}(j) \rangle)$ nominates a rule that creates $j : l$ after $jm(c1, 21_1)$ or consumes $j : l$ before. If $j = 21\text{-}1$, the edges is created by $sm(c1, c3, 2_1)$, but $\rho(\langle jm(c1, 21_1), \hat{L}(21\text{-}1) \rangle) = sm(c1, c3, 2_1)$ implies $jm(c1, 21_1) \nearrow_\rho sm(c1, c3, 2_1)$. Due to the dependency via $m21$, $sm(c1, c3, 2_1) < jm(c1, 21_1)$ and so $sm(c1, c3, 2_1) \nearrow_\rho jm(c1, 21_1)$. Since the set of causes of $jm(c1, 21_1)$ has a cyclic dependency, $sm(c1, c3, 2_1)$ is not a valid choice for ρ. Instead, $\rho(\langle jm(c1, 21_1), \hat{L}(21\text{-}1) \rangle) = lm(21\text{-}1)$ is the rule deleting the edge, thus enabling $jm(c1, 21_1)$, and hence $lm(21\text{-}1) <_\rho jm(c1, 21_1)$.

If $j = 21\text{-}1_2 \in jm(c1, 21_2)^\bullet$, then $\rho(\langle jm(c1, 21_1), \hat{L}(21\text{-}1_2) \rangle) = jm(c1, 21_2)$, which implies $jm(c1, 21_1) \nearrow_\rho jm(c1, 21_2)$. Dually, $jm(c1, 21_2) \nearrow_\rho jm(c1, 21_1)$, so both occurrences are in conflict because each creates an edge violating the other's constraint. Hence they cannot both be part of the same derivation.

If $j = 21\text{-}1_1 \in jm(c1, 21_1)^\bullet$ with $\rho(\langle jm(c1, 21_1), \hat{L}(21\text{-}1_1) \rangle) = jm(c1, 21_1)$, this is a self-disabling constraint with no extra dependencies or conflicts.

A similar analysis finds valid assignments for all rules of $\mathcal{O}$ from Example 2, so we can define a conditional occurrence grammar $\mathcal{CO}$ with the same rule occurrences, but stronger dependencies and conflicts.

As stated at the end of Sect. 2, we assume for each conditional grammar that its NACs subsume the dangling condition. This allows to show the following.

Proposition 3 (executability). *A conditional occurrence grammar $\mathcal{CO} = \langle TG, G_{in}, P, \pi, \Phi \rangle$ is executable, i.e., for each rule $p \in P$ there exists a derivation $G_{in} \Rightarrow^* G \xrightarrow{p,m} H$ in $\mathcal{CO}$.*

Proof. Let $p : \langle L_p, t_L \rangle \xleftarrow{l} \langle K_p, t_K \rangle \xrightarrow{r} \langle R_p, t_R \rangle$. By Definition 6.1 we know that the typing morphism t_L is injective.

Since we also know, by hypothesis, that the NACs subsume the dangling condition, by the "gluing conditions" of the DPO approach it is sufficient to show that there is a derivation $G_{in} \Rightarrow^* G$ such that (†) $t_L(L_p) \subseteq G$ and (‡) t_L satisfies the NAC $\Phi(p)$.

By Definition 6.2 there is an assignment ρ such that $\lfloor p \rfloor_\rho$ is finite and $(\nearrow_\rho)_{\lfloor p \rfloor_\rho}$ is acyclic. Let $q_1, \ldots, q_k, q_{k+1} = p$ be an arbitrary linearization of a finite set of rules P such that $\lfloor p \rfloor_\rho \subseteq P$ and $\lfloor P \rfloor_\rho = P$, compatible with $\nearrow_\rho$ (i.e., $q_i \nearrow_\rho q_j$ implies $i < j$ for all $i, j \in \{1, \ldots, k+1\}$). Then we show by induction on k that $G_{in} \xrightarrow{q_1, m_1} G_1 \xrightarrow{q_2, m_2} G_2 \ldots \xrightarrow{q_k, m_k} G_k$ is a derivation such that (†) $t_L(L_p) \subseteq G_k$ and (‡) t_L satisfies the NAC $\Phi(p)$.

($k = 0$). In this case $\lfloor p \rfloor = \{p\}$, thus each item $x \in t_L(L_p)$ must belong to G_{in} (otherwise for the only rule (by Definition 6.3) p_x creating x we would have $p_x \nearrow_\rho p$, by Definition 1). Therefore we have (†) $t_L(L_p) \subseteq G_{in}$. Furthermore, for each $\langle p, \hat{L}_p \rangle \in \Phi_{CO}$, let $\hat{p} = \rho(\langle p, \hat{L}_p \rangle)$. Then it cannot be the case that $\hat{p} \in (\hat{L}_p \setminus L_p)^\bullet$, otherwise $\hat{p} <_\rho p$. Thus $\hat{p} \in {}^\bullet(\hat{L}_p \setminus L_p)$, which means that there is at least one item $x \in \hat{L}_p \setminus L_p$ which does not belong to G_{in}, implying that t_L satisfies the constraint $L_p \to \hat{L}_p$. As this holds for all constraints of p, we have that (‡) t_L satisfies the NAC $\Phi(p)$.

($k \to k+1$). Let $q_1, \ldots, q_k, q_{k+1}, q_{k+2} = p$ be a linearization of $P \supseteq \lfloor p \rfloor_\rho$ with $\lfloor P \rfloor_\rho = P$ compatible with $\nearrow_\rho$. Clearly $\lfloor q_{k+1} \rfloor_\rho \subseteq P \setminus \{p\}$, and therefore by induction hypothesis we know that $G_{in} \xrightarrow{q_1, m_1} G_1 \xrightarrow{q_2, m_2} G_2 \ldots \xrightarrow{q_k, m_k} G_k$ is a derivation, that $t_{L_{k+1}}(L_{k+1}) \subseteq G_k$, and that $t_{L_{k+1}}$ satisfies the NAC $\Phi(q_{k+1})$.

Thus we can apply q_{k+1} obtaining the transformation $G_k \xrightarrow{q_{k+1}, t_{L_{k+1}}} G_{k+1}$. Let us show that (†) $t_L(L_p) \subseteq G_{k+1}$ by contradiction. Otherwise, there is an $x \in t_L(L_p)$ such that $x \notin G_{k+1}$. Since all the rules in $\lfloor p \rfloor$ are applied in the derivation (including the one generating x, if any, or else $x \in G_{in}$), there must be a rule q_i consuming x in the derivation (i.e. such that $x \in t_{L_i}(L_i \setminus K_i)$). But in this case $p \nearrow_\rho q_i$, thus the linearization would not be compatible with $\nearrow_\rho$. Concerning the satisfaction of NAC (‡), let $\langle p, \hat{L}_p \rangle \in \Phi_{CO}$ and let $\hat{p} = \rho(\langle p, \hat{L}_p \rangle)$. If $\hat{p} \in {}^\bullet(\hat{L}_p \setminus L_p)$ then $p \nearrow_\rho \hat{p}$ and the constraint is satisfied because $\hat{L}_p$ is not yet complete when p is applied. If $\hat{p} \in (\hat{L}_p \setminus L_p)^\bullet$ then $\hat{p} \nearrow_\rho p$, implying that $\hat{p}$ was applied in the derivation before p, consuming at least one item of the constraint, that is therefore satisfied by p. $\qquad\square$

5 Unfolding Conditional Graph Grammars

This section introduces the unfolding construction which, applied to a conditional grammar $\mathcal{CG}$, produces a conditional occurrence grammar $\mathcal{U}_{\mathcal{CG}}$ describing the behaviour of $\mathcal{CG}$. The unfolding is equipped with a morphism $u_{\mathcal{CG}}$ to the original grammar $\mathcal{CG}$ which allows to see rules in $\mathcal{U}_{\mathcal{CG}}$ as instances of rule applications in $\mathcal{CG}$, and items of the type graph of $\mathcal{U}_{\mathcal{CG}}$ as instances of items of the type graph of $\mathcal{CG}$.

Starting from the initial graph of the grammar, we apply in all possible ways its rules, and record in the unfolding each redex and each new graph item generated, both enriched with their corresponding causal history. In order to account for the satisfaction of NACs, for each rule q we check that for all occurrences of its negative constraints $\hat{L}_q$ present in one of its histories, there is another rule r deleting part of $\hat{L}_q \setminus L_q$. In that case we know that there exists a history in which the constraint is no longer present when q is applied.

We introduce conditional concurrent graphs as the subgraphs of the type graph coverable in the conditional occurrence grammar.

Definition 7 (conditional concurrent graph). *Let* $CO = \langle TG, G_{in}, P, \pi, \Phi \rangle$ *be a conditional occurrence grammar. A subgraph G of TG is called* conditional concurrent *if there is an assignment ρ such that*

1. $\neg(x <_\rho y)$ *for all* $x, y \in G$.
2. $\lfloor G \rfloor_\rho$ *is finite*
3. $\nearrow_\rho$ *is acyclic on* $\lfloor G \rfloor_\rho$.

Proposition 4 (conditional concurrent graphs are coverable). *A graph $G \subseteq TG$ is conditional concurrent if and only if it is coverable in CO.*

Proof. Given a derivation $s = (G_{in} = G_0 \xrightarrow{p_1, m_1} \cdots \xrightarrow{p_n, m_n} G_n = G)$ in CO, it is also a derivation in the underlying unconditional occurrence grammar O. That means, G is a concurrent graph in the sense of Definition 3. Let ρ be defined for all p_i in the derivation and any constraint $\hat{L}_{p_i}$ of p_i as follows. Since s is a conditional derivation, we know that $\hat{L}_{p_i} \not\subseteq G_{i-1}$. Then if $\hat{L}_{p_i} \subseteq G_k$ for $k < i-1$, there must exist a rule $p_j \in (\hat{L}_{p_i} \setminus L_{p_i})^\bullet$ with $k < j < i$. In this case we set $\rho(\langle p_i, \hat{L}_{p_i} \rangle) = p_j$, which induces the new dependency $p_j <_\rho p_i$. Otherwise, we set $\rho(\langle p_i, \hat{L}_{p_i} \rangle) = q$, where q is an arbitrary rule in $^\bullet(\hat{L}_{p_i} \setminus L_{p_i})$. Note that in this case $q \notin \{p_1, \ldots p_i\}$, and the new induced dependency is $p_i \nearrow_\rho q$. This defines ρ on all constraints in $\bigcup_{i \in \{1,\ldots,n\}} \Phi(p_i)$: it can be shown that ρ can be extended to all constraints in $\Phi(CO)$ without introducing cycles of dependencies among rules in $\{p_1, \ldots, p_n\}$. Then we can easily check that G is conditional concurrent w.r.t. ρ. In fact,

1. no new dependencies are added between items of G, i.e., $\neg(x <_\rho y)$ for all $x, y \in G$;
2. $\lfloor G \rfloor_\rho \subseteq \{p_1, \ldots, p_n\}$, thus it is finite;
3. the new dependencies cannot make $\nearrow_\rho$ cyclic on $\lfloor G \rfloor_\rho$, as all of them are consistent with the ordering of rules in the derivation.

Furthermore, by Definition 7 it is obvious that conditional concurrent graphs are closed under inclusion, thus all coverable graphs are conditional concurrent.

Vice versa, if G is a conditional concurrent graph w.r.t. assignment ρ, let $p_1, \ldots, p_n$ be a linearization of $\lfloor G \rfloor_\rho$ compatible with $\nearrow_\rho$. Then following the same outline of the proof of Proposition 3, it is possible to prove by induction on n that there is a derivation $s = (G_{in} = G_0 \xrightarrow{p_1, m_1} \cdots \xrightarrow{p_n, m_n} G_n)$ in CO such that $G \subseteq G_n$. $\qquad \square$

Another basic ingredient of the unfolding is the gluing operation, that we borrow literally from [1]. It can be interpreted as a "partial application" of a rule to a given match, in the sense that it generates the new items as specified by the rule (i.e., items of right-hand side not in the interface), but items that should have been deleted are not affected: intuitively, this is because such items may still be used by another rule in the nondeterministic unfolding. In the following we assume that for each rule name q its associated rule is $L_q \leftarrow K_q \rightarrow R_q$, where the injections l_q and r_q are inclusions (and not generic injective morphisms).

Definition 8 (gluing). *Let q be a rule, G a graph and $m : L_q \rightarrow G$ a graph morphism. We define, for any given symbol $*$, the gluing of G and R_q along K_q, according to m and marked by $*$, denoted by $glue_*(q, m, G)$ as the graph $\langle N, E, s, t \rangle$, where:*

$$N = N_G \cup m_*(N_{R_q}) \qquad E = E_G \cup m_*(E_{R_q})$$

with m_* defined by: $m_*(x) = m(x)$ if $x \in K_q$ and $m_*(x) = \langle x, * \rangle$ otherwise. The source and target functions s and t, and the typing are inherited from G and R_q.

The gluing operation keeps unchanged the identity of the items already in G, and records in each newly added item from R_q the given symbol $*$. We remark that the gluing, as just defined, is a concrete deterministic definition of the pushout of the arrows $G \xleftarrow{m} L_q \xleftarrow{l_q} K_q$ and $K_q \xrightarrow{r_q} R_q$.

In the construction that follows, if q is a rule with NAC $\Phi(q)$ and $\langle q, m \rangle$ is a redex in the type graph TG of a conditional occurrence grammar, then we let $\Phi_{TG}(\langle q, m \rangle)$ denote the set of occurrences of q's constraints in TG relevant to m. That means, $\Phi_{TG}(\langle q, m \rangle)$ is the set of all graphs $k(\hat{L}_q)$ such that there exists a constraint $n : L_q \rightarrow \hat{L}_q \in \Phi(q)$ and a mono $k : \hat{L}_q \rightarrow TG$ with $m = k \circ n$.

Now the unfolding of a conditional grammar $\mathcal{CG} = \langle TG, G_{in}, P, \pi, \Phi \rangle$ is defined as follows. For each $n \in \mathbb{N}$, we construct a partial unfolding $\mathcal{U}(\mathcal{CG})^{(n)} = \langle \mathcal{U}^{(n)}, u^{(n)} \rangle$, where $\mathcal{U}^{(n)} = \langle TG^{(n)}, P^{(n)}, \pi^{(n)}, \Phi^{(n)} \rangle$ is a conditional occurrence grammar and $u^{(n)} = \langle u_{TG}^{(n)}, u_P^{(n)} \rangle : \mathcal{U}(\mathcal{CG})^{(n)} \rightarrow \mathcal{CG}$ a CGG morphism. Intuitively, the occurrence grammar generated at level n contains all possible computations of the grammar with "causal depth" at most n.

- **(n = 0)** $\langle \mathcal{U}^{(0)}, u^{(0)} \rangle$ is defined as $\mathcal{U}^{(0)} = \langle G_{in}, \emptyset, \emptyset, \emptyset \rangle$ and $u_{TG}^{(0)} = t_{G_{in}}$.
- **(n → n + 1)** Given $\mathcal{U}(\mathcal{CG})^{(n)}$, the partial unfolding $\mathcal{U}(\mathcal{CG})^{(n+1)}$ is obtained by extending it with all the compatible applications of rules of P to concurrent subgraphs of the type graph of $\mathcal{U}^{(n)}$. Given a rule $q \in P$ and match $m : L_q \rightarrow \langle TG^{(n)}, u_{TG}^{(n)} \rangle$, redex $\langle q, m \rangle$ *is an occurrence of q compatible with $\mathcal{U}(\mathcal{CG})^{(n)}$* if there exists an assignment ρ such that
  1. graph $m(L_q) \subseteq TG^{(n)}$ is conditional concurrent w.r.t. ρ, and
  2. for each constraint occurrence $\hat{L}_q \in \Phi_{TG^{(n)}}(\langle q, m \rangle)$, there exists a rule
  $$r \in (\hat{L}_q \setminus L_q)^{\bullet} \cap \lfloor m(L_q) \rfloor_\rho \text{ or a rule } r \in {}^{\bullet}(\hat{L}_q \setminus L_q) \setminus \lfloor m(L_q) \rfloor_\rho.$$
  Let P_{n+1} be the set of all $\langle q_i, m_i \rangle$, for $i = 1, \ldots, k$, such that $q_i \in P$ and $\langle q_i, m_i \rangle$ is an occurrence of q_i compatible with $\mathcal{U}(\mathcal{CG})^{(n)}$. Then, $\mathcal{U}(\mathcal{CG})^{(n+1)}$ is given by

- $TG^{(n+1)}$ is the consecutive gluing of $TG^{(n)}$ with $R_{q_1}, \ldots, R_{q_k}$ along $K_{q_1}, \ldots, K_{q_k}$ respectively. Formally, $TG^{(n+1)} = TG_k$ where $TG_0 = TG^{(n)}$, and $TG_i = glue_{\langle q_i, m_i \rangle}(q_i, m_i, TG_{i-1})$ for $i \in \{1, k\}$. Note that the result is independent of the order of rule matches applied.
- Morphism $u_{TG}^{(n)}$ extends canonically to $u_{TG}^{(n+1)} : TG^{(n+1)} \to TG$.
- $P^{(n+1)} = P^{(n)} \cup P_{n+1}$.
- $u_P^{(n+1)} = u_P^{(n)} \cup u_{P,n+1}$ where $u_{P,n+1}(\langle q, m \rangle) = q$.
- $\pi^{(n+1)}(\langle q, m \rangle)$ coincides with $\pi(q)$ except for the typing.
- $\Phi^{(n+1)}(\langle q, m \rangle) = \Phi_{TG^{(n+1)}}(\langle q, m \rangle)$ is the set of all occurrences of q's constraints relevant to m in $TG^{(n+1)}$.

Note that if a rule is applied twice (also in different steps) at the same match, the generated items are the same and thus they appear only once in the unfolding. By construction it is evident that $\mathcal{U}(\mathcal{CG})^{(n)} \subseteq \mathcal{U}(\mathcal{CG})^{(n+1)}$, componentwise.

Definition 9 (unfolding). *The unfolding $\mathcal{U}(\mathcal{CG}) = \langle \mathcal{U}_{\mathcal{CG}}, u_{\mathcal{CG}} \rangle$ of the grammar $\mathcal{CG}$ is defined as $\bigcup_n \mathcal{U}(\mathcal{CG})^{(n)}$, where union is applied componentwise.*

Example 3 provides part of the unfolding of the grammar of Example 1. Step $(n = 1)$ adds $pc(c2)$ and corresponding occurrences $pc(c1)$ and $pc(c3)$, creating 3 Server nodes. Step $(n = 2)$ creates two occurrences of sm per Server, each creating a meeting node. Step $(n = 3)$ adds 2 occurrences of lm per meeting, allowing participants to leave. This enables 4 occurrences of jm per meeting in $(n = 4)$ which in turn enables the remaining 4 instances of lm per meeting in step $(n = 5)$ and then one occurrence of tm per meeting in step $(n = 6)$. The folding morphism u is defined by mapping rules by name and types as $u_{TG}(x : X) = X$.

We conclude by showing that $\mathcal{U}_{\mathcal{CG}}$ is a conditional occurrence grammar, and that $u_{\mathcal{CG}} : \mathcal{U}_{\mathcal{CG}} \to \mathcal{CG}$ is a conditional grammar morphism. From the proof it follows that the same holds for all finite approximations $\mathcal{U}^{(n)}(\mathcal{CG})$.

Proposition 5 (conditional unfolding). *$\mathcal{U}_{\mathcal{CG}}$ is a conditional occurrence grammar and $u_{\mathcal{CG}} : \mathcal{U}_{\mathcal{CG}} \to \mathcal{CG}$ is a conditional grammar morphism.*

Proof. Condition 1 of Definition 6 is satisfied by construction: elements of G_{in} and those created by each rule occurrence are added as fresh elements to the type graph, and all rules are injective by definition. Further, if we drop the NACs Φ from the unfolding as constructed above, we obtain an unconditional occurrence grammar. This ensures condition 3 in Definition 6.

It remains to show that for all $q \in P_{\mathcal{U}_{\mathcal{CG}}}$, there exists an assignment ρ such that $\lfloor q \rfloor_\rho$ is finite and $(\nearrow_\rho)_{\lfloor q \rfloor_\rho}$ is acyclic. For $n = 0$ the set of rules is empty, so the condition trivially holds. For the inductive step, note that $P^{(n+1)} = P^{(n)} \cup P_{(n+1)}$. We consider the two cases separately. If $p = \langle q, m \rangle \in P_{(n+1)}$, thus p was added in step $(n + 1)$, we know by construction that $m(L_q) \subseteq TG^{(n)}$ is a conditional concurrent subgraph w.r.t. an assigment ρ'. We can extend ρ' to an assignment $\rho'' : \Phi(\mathcal{U}^{(n+1)}) \to P^{(n+1)}$ as follows:

$$\rho''(\langle \hat{L}, q' \rangle) = \begin{cases} \rho'(\langle \hat{L}, q' \rangle) & \text{if } \langle \hat{L}, q' \rangle \in \Phi(\mathcal{U}^{(n)}) \\ r & \text{if } p = q' \text{ and } r \text{ is determined as in point} \\ & \text{2 of construction step } (n \to n+1) \\ \text{any } r \in {}^{\bullet}(\hat{L} \setminus L_{q'}) & \text{if } q' \in P_{(n+1)} \setminus \{p\} \end{cases}$$

Then $\lfloor p \rfloor_{\rho''}$ is finite because so is $\lfloor m(L_q) \rfloor_{\rho'}$, and $(\nearrow_{\rho''})\lfloor p \rfloor_{\rho''}$ is acyclic, because new relevant dependencies are added only in the second case above: if $\rho''(\langle \hat{L}_q, p \rangle) = r \in (\hat{L}_q \setminus L_q)^{\bullet} \cap \lfloor m(L_q) \rfloor_{\rho}$ then $r <_{\rho''} p$, but r is already a cause of p; otherwise $r \in {}^{\bullet}(\hat{L}_q \setminus L_q) \setminus \lfloor m(L_q) \rfloor_{\rho}$ and r is not among the causes of p.

If instead $p = \langle q, m \rangle \in P^{(n)}$, by inductive hypothesis we know that there is an assignment for it satisfying the required condition and defined on $\Phi(\mathcal{U}^{(n)})$. We can easily extend it to the constraints in $\Phi(\mathcal{U}^{(n+1)}) \setminus \Phi(\mathcal{U}^{(n)})$, which are generated at step $(n+1)$, by mapping them to one of the rules generating the constraint. This does not affect neither the set of causes of p nor the dependencies among them. Concerning $u_{\mathcal{CG}} : \mathcal{U}_{\mathcal{CG}} \to \mathcal{CG}$, the fact that it is a conditional grammar morphism follows by the fact that so are all its finite approximations $u^{(n)} : \mathcal{U}^{(n)} \to \mathcal{CG}$, which in turn is pretty obvious because the conditions of Definition 4 are enforced by construction at each step. $\qquad \square$

It is possible to show that the unfolding construction applied to a conditional occurrence grammar yields a grammar which is isomorphic to the original one.

6   Incremental NACs

With general NACs allowing complex forbidden patterns, for each constraint occurrence there can be a number of choices for the assignment ρ among the rules creating or destroying parts of the occurrence. These choices multiply as the assignment is defined for all occurrences of all constraints of all rules in the occurrence grammar. This complexity is particularly worrying in the unfolding construction, where a suitable assignment has to be determined for each match of each rule to be applied.

We consider the class of incremental NACs for which such choices can be made more canonically, reducing the complexity. In [13] these are defined categorically. While working with concrete typed graphs, it is sufficient here to say that they only allow constraints containing either isolated negative nodes or edges attached to nodes in the left-hand side with their opposite source or target nodes. E.g., in the grammar of Example 1, all NACs are incremental apart from the one of rule $sm(c1, c2, i)$. When defining an assignment for its occurrences, e.g., $\rho(\langle sm(c1, c3, 2_1), \hat{L}(21\text{-}1_1, 21\text{-}3_1) \rangle) \in {}^{\bullet}\hat{L}(21\text{-}1_1, 21\text{-}3_1)$ can map to either one of $jm(c1, 21_1)$ or $jm(c3, 21_1)$ each of which creates one of the edges. Hence the choice is not canonical.

Instead, occurrences of incremental NACs cannot be created or destroyed in two or more independent steps. For example, a constraint $\hat{L}$ containing an outgoing edge and its target node can be established by first creating the node and then the edge, but not vice versa. That means, in any occurrence grammar and for any assignment ρ for rule q the rules in $^\bullet(\hat{L} \setminus L) \cap \lfloor q \rfloor_\rho$ as well as $(\hat{L} \setminus L)^\bullet \cap \lfloor q \rfloor_\rho$ are linearly ordered by $\nearrow_\rho$.

Given an occurrence grammar and rule p we can order assignments ρ satisfying the clause (1) of Definition 6 by inclusion of the relation $\nearrow_\rho$ induced as $\rho' \leq \rho$ iff $(\nearrow_{\rho'})^+ \subseteq (\nearrow_\rho)^+$.

Since $\leq$ is a partial order on assignments, this gives us a notion of *minimal assignment*, leading to an asymmetric conflict relation that allows maximal concurrency. We can "improve" an assignment ρ for a rule p (make the order more concurrent) by picking a rule $q \in \lfloor L_p \rfloor_\rho$ and, for each constraint occurrence $\hat{L}_q$:

- If $\rho(\langle q, \hat{L}_q \rangle) = r \in {}^\bullet(\hat{L}_q \setminus L_q)$ let $\rho'(\langle q, \hat{L}_q \rangle) = r'$ be such that $r \nearrow_\rho r'$ and r' is maximal within ${}^\bullet(\hat{L}_q \setminus L_q)$ with that property
- if $\rho(\langle q, \hat{L}_q \rangle) = r \in (\hat{L}_q \setminus L_q)^\bullet$ let $\rho'(\langle q, \hat{L}_q \rangle) = r'$ be such that $r' \nearrow_\rho r$ and r' is minimal within $(\hat{L}_q \setminus L_q)^\bullet$ with that property.

Intuitively we maximise the freedom of q by defining its limits as widely as possible. This leads to a weakening of the relation, such that after repeated application of such improvement step we arrive at an assignment that is minimal w.r.t. the order defined above.

In general, such a minimum is not unique, but for incremental NACs it is easy to see that there is a unique minimal or maximal r' to be chosen in the improvement step above, leading to a single most concurrent assignment for each rule p. That means, in step $n + 1$ of the unfolding, we can choose this minimal assignment for every new rule q', avoiding the need to try and discard alternatives that do not lead to acyclic histories. In fact, since the minimal assignment represents the weakest relation, if there is an assignment ρ leading to an acyclic $\nearrow_\rho$ on $\lfloor m(L_q) \rfloor_\rho$, then the minimal assignment also has that property.

7   Conclusion

Aiming to generalise the theory of unfolding of both typed graph grammars and P/T nets with read and inhibitor arcs, in this paper we present occurrence grammars and unfolding for conditional graph grammars and prove their fundamental relationships with each other and the derivations in the grammar. This establishes conditional unfolding as a sound semantic model of graph grammars with negative application conditions. We believe that the proof that unfoldings give rise to a co-reflection, which is often part of this theory, requires a restriction to incremental NACs.

The constructions in this paper can be generalised in several ways, which we would also like to explore in the future. First, like in the classical case, we are limited to consuming rules. For unconditional rules this ensures that the same rule cannot be applied twice at the same match, but for rules with NACs the same can be achieved by so-called self-disabling rules that generate their own NACs. However, the current formulation of unfolding, based on a restriction of the construction for the unconditional case, leads us to identify recurrent rule matches with self-disabling NACs. A stronger notion of rule occurrence, taking account of the history also in terms of the NACs created and destroyed, is required to address this problem.

The work presented here is part of a wider project of using graph grammars to specify and solve graph-based combinatorial optimisation problem. In this context, unfoldings provide an efficient form of breadth-first search through the state space of the grammar which, if the grammar is terminating, will yield a finite set of output graphs. Apart from exploring the advantages of incremental NACs we will consider attributed graph grammars to model combinatorial problem with data.

References

1. Baldan, P., Corradini, A., Montanari, U., Ribeiro, L.: Unfolding semantics of graph transformation. Inf. Comput. **205**(5), 733–782 (2007)
2. Baldan, P., Corradini, A., Heindel, T., König, B., Sobociński, P.: Unfolding grammars in adhesive categories. In: Kurz, A., Lenisa, M., Tarlecki, A. (eds.) CALCO 2009. LNCS, vol. 5728, pp. 350–366. Springer, Heidelberg (2009). https://doi.org/10.1007/978-3-642-03741-2_24
3. Baldan, P., Corradini, A., Montanari, U., Ribeiro, L.: Coreflective concurrent semantics for single-pushout graph grammars. In: Wirsing, M., Pattinson, D., Hennicker, R. (eds.) WADT 2002. LNCS, vol. 2755, pp. 165–184. Springer, Heidelberg (2003). https://doi.org/10.1007/978-3-540-40020-2_9
4. Baldan, P., König, B.: Approximating the behaviour of graph transformation systems. In: Corradini, A., Ehrig, H., Kreowski, H.-J., Rozenberg, G. (eds.) ICGT 2002. LNCS, vol. 2505, pp. 14–29. Springer, Heidelberg (2002). https://doi.org/10.1007/3-540-45832-8_4
5. Ehrig, H., Pfender, M., Schneider, H.J.: Graph-grammars: an algebraic approach. In: 14th Annual Symposium on Switching and Automata Theory, pp. 167–180. IEEE Computer Society (1973)
6. Ehrig, H., Ehrig, K., Prange, U., Taentzer, G.: Fundamentals of Algebraic Graph Transformation. Monographs in Theoretical Computer Science. An EATCS Series. MTCSAES. Springer, Heidelberg (2006). https://doi.org/10.1007/3-540-31188-2
7. Baldan, P., Corradini, A., König, B.: Verifying finite-state graph grammars: an unfolding-based approach. In: Gardner, P., Yoshida, N. (eds.) CONCUR 2004. LNCS, vol. 3170, pp. 83–98. Springer, Heidelberg (2004). https://doi.org/10.1007/978-3-540-28644-8_6
8. Qayum, F., Heckel, R.: Search-based refactoring using unfolding of graph transformation systems. ECEASST **38** (2011)
9. Habel, A., Heckel, R., Taentzer, G.: Graph grammars with negative application conditions. Fundam. Inform. **26**(3/4), 287–313 (1996)
10. Ehrig, H., Golas, U., Habel, A., Lambers, L., Orejas, F.: $\mathcal{M}$-adhesive transformation systems with nested application conditions. Part 1: parallelism, concurrency and amalgamation. MSCS **24**(4) (2014)
11. Ehrig, H., Golas, U., Habel, A., Lambers, L., Orejas, F.: $\mathcal{M}$-adhesive transformation systems with nested application conditions. Part 2: embedding, critical pairs and local confluence. Fundam. Inform. **118**(1–2), 35–63 (2012)

12. Hermann, F., Corradini, A., Ehrig, H.: Analysis of permutation equivalence in - adhesive transformation systems with negative application conditions. MSCS **24**(4) (2014)
13. Corradini, A., Heckel, R.: Canonical derivations with negative application conditions. In: Giese, H., König, B. (eds.) ICGT 2014. LNCS, vol. 8571, pp. 207–221. Springer, Cham (2014). https://doi.org/10.1007/978-3-319-09108-2_14
14. Baldan, P., Busi, N., Corradini, A., Pinna, G.M.: Domain and event structure semantics for Petri nets with read and inhibitor arcs. Theor. Comput. Sci. **323**(1–3), 129–189 (2004)
15. Corradini, A., Montanari, U., Rossi, F.: Graph processes. Fundam. Inform. **26**(3/4), 241–265 (1996)

Two-Level Reasoning About Graph Transformation Programs

Amani Makhlouf, Christian Percebois[✉], and Hanh Nhi Tran

IRIT, University of Toulouse, Toulouse, France
{Amani.Makhlouf,Christian.Percebois,Hanh-Nhi.Tran}@irit.fr

Abstract. This paper presents a method for verifying graph transformation programs written in Small-t$\mathcal{ALC}$, an imperative language which allows expressing graph properties and graph transformations in $\mathcal{ALCQI}$ description logic. We aim at reasoning not only about the local effect when applying a transformation rule on a matched subgraph but also about the global impact on the whole input graph when applying a set of rules. Using $\mathcal{ALCQI}$ assertional and terminological formulae to formalize directed labeled graphs, Small-t$\mathcal{ALC}$ allows specifying local properties on individual nodes and edges as well as global properties on sets of nodes and edges. Our previous work focuses on verifying local properties of the graph. In this paper, we propose a static analyzer at terminological level that intertwines with a static analyzer at assertional level to infer global properties of the transformed graph.

Keywords: Graph transformation · Description logics ·
Static analysis · Abstract interpretation · Program verification

1 Introduction

To allow verifying the correctness of graph transformations, many works, rooted in algebraic approach for formalizing graph transformations, have introduced logic systems that are specially tailored for expressing graph properties under study (see e.g. [1–5]).

The work presented in this paper uses another approach which directly encodes graphs in an existing logic [6,7] in order to benefit the inference mechanisms provided for the chosen logic. Adopting this approach, we proposed the graph transformation language Small-t$\mathcal{ALC}$ [8] which specifies graphs with $\mathcal{ALCQI}$ description logic formulae [9] and defines transformation statements to manipulate graphs in an imperative paradigm. Transformation specifications and code are based on the same logic thus we can take advantage of a Hoare-like calculus and also of proven program verification techniques to reason about the correctness of graph transformations.

Small-t$\mathcal{ALC}$ graphs are directed and labeled. A graph consists of nodes representing individuals and edges representing relations between individuals. A node can be labeled to express that it belongs to the concept denoted by the node's

© Springer Nature Switzerland AG 2019
E. Guerra and F. Orejas (Eds.): ICGT 2019, LNCS 11629, pp. 111–127, 2019.
https://doi.org/10.1007/978-3-030-23611-3_7

label; the label of an edge denotes the role of the relation represented by the edge. Graph properties can be specified by $\mathcal{ALCQI}$ assertional axioms ($ABox$) about nodes and edges and by terminological axioms ($TBox$) about set of nodes.

A Small-t$\mathcal{ALC}$ program consists of a set of transformation rules. Each rule comprises a precondition specifying the matching constraints of the rule on a host graph, a code consisting of transformation statements and a postcondition specifying the properties of the graph yielded from the rule's application. Both rule's specifications and code are formalized at $ABox$ level. In [7,10], we developed tools to formally verify the correctness of each transformation rule using Hoare logic. However, our previous works allow verifying only a plain set of rules, not the correctness of a whole transformation program. Moreover, using only assertional formulae to specify graph properties, we could analyze only local properties on individual nodes and edges.

We now extend the approach to reason not only about the local effect when applying a transformation rule on a matched subgraph but also about the global impact on the whole input graph when applying a set of rules. For this purpose, first we exploit both $\mathcal{ALCQI}$ assertional formulae ($ABox$) and terminological formulae ($TBox$) to formalize directed labeled graphs, and thus allow specifying respectively local properties as well as global properties. We then propose a static analyzer at terminological level that intertwines with a static analyzer at assertional level to infer global properties of the transformed graph. Rules verification at $ABox$ level was presented in [10]. The focus of this paper is the $TBox$ analyzer for transformation programs.

We introduces our graph transformation language Small-t$\mathcal{ALC}$ in Sect. 2 and present in Sect. 3 the main idea of two-levels reasoning about Small-t$\mathcal{ALC}$ programs by exploiting the $ABox$ and $TBox$ components of $\mathcal{ALCQI}$. In Sect. 4 we explain how to infer, by abstract interpretation, $TBox$ global properties from $ABox$ statements. The relation between $ABox$ and $TBox$ verifications is studied in Sect. 5. Section 6 shows that some monadic second-order properties can be expressed by Small-t$\mathcal{ALC}$ $TBox$ assertions too. We finally provide some discussions on related work in Sect. 7 and wrap up the paper with a conclusion including further work in Sect. 8.

2   The Small-t$\mathcal{ALC}$ Language

Small-t$\mathcal{ALC}$ [8] is an imperative graph transformation language based on the description logic $\mathcal{ALCQI}$ [9]. The distinctive characteristic of this graph transformation language is the tight integration of logical aspects with the intended execution mechanism, with the overall aim to obtain a decidable calculus for reasoning about program correctness in a pre-/post-condition style.

2.1   Logic Foundation

$\mathcal{ALCQI}$ represents knowledge at two levels: $TBox$ introduces the terminology, i.e., the vocabulary of an application domain, while $ABox$ contains assertions about named individuals in terms of this vocabulary. The vocabulary consists of concepts, which denote sets of individuals, and roles, which denote binary relationships between individuals. An interpretation $\mathcal{I}$ that is used to define the semantics of DLs comprises a non-empty set $\Delta^{\mathcal{I}}$ called the interpretation domain and an interpretation function $\cdot^{\mathcal{I}}$. The interpretation function assigns an element $i^{\mathcal{I}} \in \Delta^{\mathcal{I}}$ to each individual i of the *ABox*, a subset of individuals $C^{\mathcal{I}} \in \Delta^{\mathcal{I}}$ to each concept C of the *TBox*, and a subset of ordered pairs of individuals $r^{\mathcal{I}} \in \Delta^{\mathcal{I}} \times \Delta^{\mathcal{I}}$ to every role r of the *TBox*.

Let C be a concept, x and y be individuals, and r be a role. If x belongs to the concept C, then x is called C-type. If x is r-related to y, then y is called a r-successor of x. $\mathcal{ALCQI}$ provides concept constructors to build more complex concepts as given in Table 1.

Table 1. $\mathcal{ALCQI}$ concept constructors

Name	Syntax	Semantics
top	$\top$	$\Delta^{\mathcal{I}}$
bottom	$\bot$	$\emptyset$
negation	$\neg C$	$\Delta^{\mathcal{I}} \backslash C^{\mathcal{I}}$
conjunction	$C \cap D$	$C^{\mathcal{I}} \cap D^{\mathcal{I}}$
disjunction	$C \cup D$	$C^{\mathcal{I}} \cup D^{\mathcal{I}}$
existential restriction	$\exists\, r\; C$	$\{x \in \Delta^{\mathcal{I}} \mid \exists y, (x,y) \in r^{\mathcal{I}} \wedge y \in C^{\mathcal{I}}\}$
universal restriction	$\forall\, r\; C$	$\{x \in \Delta^{\mathcal{I}} \mid \forall y, (x,y) \in r^{\mathcal{I}} \Rightarrow y \in C^{\mathcal{I}}\}$
at-most restriction	$\leq n\, r\; C$	$\{x \in \Delta^{\mathcal{I}} \mid \mid(x,y) \in r^{\mathcal{I}} \wedge y \in C^{\mathcal{I}}\mid \leq n\}$
at-least restriction	$\geq n\, r\; C$	$\{x \in \Delta^{\mathcal{I}} \mid \mid(x,y) \in r^{\mathcal{I}} \wedge y \in C^{\mathcal{I}}\mid \geq n\}$
equality restriction	$= n\, r\; C$	$\{x \in \Delta^{\mathcal{I}} \mid \mid(x,y) \in r^{\mathcal{I}} \wedge y \in C^{\mathcal{I}}\mid = n\}$
inverse role	r^{-1}	$\{(y,x) \mid (x,y) \in r^{\mathcal{I}}\}$

$(\exists\, r\; C)$ describes the set of individuals having at least a r-successor which is C-type. $(\forall\, r\; C)$ presents the set of individuals whose all r-successors are C-type. $(\leq n\, r\; C)$ and $(\geq n\, r\; C)$ are qualified number restrictions expressing that an individual has at most (respectively at least) n r-successors which are C-type.

2.2   Small-t$\mathcal{ALC}$ Graphs

An interpretation $\mathcal{I}$ can be drawn as a directed labeled graph [11] where *TBox* represents concepts and roles respectively as nodes labels and edges labels, and *ABox* specifies individuals and binary relations between them respectively as graph nodes and graph edges.

In Small-t$\mathcal{ALC}$, concept assertions $(i : C)$ express that an individual i is C-type, i.e. the node i is labeled with C (C-node). Role assertions in the form $(i\; r\; j)$ express that an individual i is connected by the role r to the individual j i.e. the edge (i,j) is labeled with r (r-edge). By combining concept assertions and role assertions, *ABox* formulae are made up and used to specify properties on named graph nodes and edges. Figure 1 depicts a graph having two *A*-nodes $a1$, $a2$ and two *B*-nodes $b1$, $b2$. $b1$ is a r-successor of $a2$ and $b2$ is a r-successor of $a1$. There are also two anonymous *C*-type nodes which are r-successors of $a1$, thus $a1$ belongs to the concept which has at least 2 *C*-nodes as r-successors.

In the rest of the paper, we call *AFact* an *ABox* assertion, and *AFormula* an *ABox* formula.

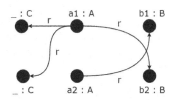

Fig. 1. A graph satisfying the *AFormula* $(a1 : A) \wedge (a2 : A) \wedge (b1 : B) \wedge (b2 : B) \wedge (a1 \ r \ b2) \wedge (a2 \ r \ b1) \wedge (a1 : (\leq 2 \ r \ C))$

TBox axioms use general concept inclusions (GCI) [12] to express properties concerning concepts. *TBox* axioms are so-called *TFacts* in Small-t$\mathcal{ALC}$ and are of the form $C \sqsubseteq D$ or $C = D$ where C and D are concepts. An interpretation $\mathcal{I}$ is a model of $C \sqsubseteq D$ if $C^{\mathcal{I}} \subseteq D^{\mathcal{I}}$. When $C^{\mathcal{I}} \subseteq D^{\mathcal{I}}$ in every model of I, D subsumes C. Thus, *TFormulae*, which are Boolean combinations of *TFacts*, can be used now to express global graph properties on set of nodes. For example, the *TFact* $(\forall \ r^{-1} \ A) \sqsubseteq B$ expresses that r-edges outgoing from A-nodes all go towards B-nodes. The graph of Fig. 1 does not hold this property because it has two r-edges outgoing from A-nodes to C-nodes.

2.3   Small-t$\mathcal{ALC}$ Statements

Small-t$\mathcal{ALC}$ features atomic statements to add, delete or select graph nodes and edges. We have defined five atomic Small-t$\mathcal{ALC}$ statements according to the following grammar, where i and j are node variables which will be bound to the host graph's nodes during the transformation's execution, C is a concept name, r is a role name, F is an $\mathcal{ALCQI}$ *AFormula* and v is a list of node variables:

$$stmt ::= \texttt{add}(i : C) \qquad (C^{\mathcal{I}} = C^{\mathcal{I}} + \{i\})$$
$$| \quad \texttt{delete}(i : C) \quad (C^{\mathcal{I}} = C^{\mathcal{I}} - \{i\})$$
$$| \quad \texttt{add}(i \ r \ i) \qquad (r^{\mathcal{I}} = r^{\mathcal{I}} + \{i, j\})$$
$$| \quad \texttt{delete}(i \ r \ i) \quad (r^{\mathcal{I}} = r^{\mathcal{I}} - \{i, j\})$$
$$| \quad \texttt{select} \ v \ with \ F$$

Operationally, the first four Small-t$\mathcal{ALC}$ statements define new interpretations i.e. new graphs by adding and deleting individuals (nodes) and pair of individuals (edges) to and from the interpretations of concepts and roles. The interpretation function of *TBox* concepts and roles are thus evolved. In this sense, a rule operates on *AFormulae* but affects as well *TFormulae*. Note that the Small-t$\mathcal{ALC}$ statements do not add/delete individuals to/from the graph, but change their label to modify the interpretation represented by the graph.

Since concepts and roles are considered as sets of nodes and set of pairs of nodes respectively, $add(i : C)$ and $add(i\ r\ j)$ have no effects if $i \in C$ and $(i, j) \in r$ respectively. Therefore, no parallel edges with the same label are allowed. The statement $delete(i : C)$ does not remove it definitely from the graph, but excludes it from the interpretation function C^I of the indicated concept C, i.e. the node will no more be labeled with C.

An original construct is the *select* statement that non-deterministically binds node variables to nodes in the subgraph that satisfies an *AFormula*. This assignment is used to select specific nodes where the transformations are requested to occur. The remaining language constructs are conventional control structures: sequence, branching and iteration.

2.4   Small-t$\mathcal{ALC}$ Programs

A Small-t$\mathcal{ALC}$ program consists of a set of transformation rules and a *main* entry point of the program. A rule is structured into three parts: a precondition, the transformation code (a sequence of statements) and a postcondition. The pre- and postconditions of a rule are two *AFormulae* which specify respectively a source graph which can be transformed by the rule and the target graph supposed to be produced by the rule.

```
rule rename {
    pre: (a : A) ∧ (b : B) ∧ (a r b)
        delete(a r b);
        add(a s b);
    post: (a : A) ∧ (b : B) ∧ (a s b) ∧ (a ¬r b)
}
rule reverse {
    pre: (a : A) ∧ (b : B) ∧ (a s b)
        delete(a s b);
        add(b r a);
    post: (a : A) ∧ (b : B) ∧ (a ¬s b)
}
main {
    assert: (∃ r⁻¹ A) ⊆ B ∧ (∃ s⁻¹ A) = ⊥
    rename !;
    reverse !;
    assert: (∃ r⁻¹ A) = ⊥ ∧ (∃ r⁻¹ B) ⊆ A ∧ (∃ s⁻¹ A) = ⊥
}
```

Fig. 2. Small-t$\mathcal{ALC}$ program *Edges − Reversing*

Rules that are defined separately in a Small-t$\mathcal{ALC}$ program are called sequentially in the *main*. Two types of rule calls are proposed: a simple call (CALL) and an iterative call (CALL!). The first executes the code of the rule if a subgraph in the source graph matches the *ABox* precondition formula. The second executes the code of the rule as long as a subgraph matches the precondition. We can inject *TFormulae* into *main* to specify the properties of the transformed graph before (*pre-TFormula*) and after (*post-TFormula*) applications of one or many rules.

We illustrate in Fig. 2 a Small-t$\mathcal{ALC}$ transformation program which reverses all r-edges from A-nodes to B-nodes. This transformation is done in two steps: first the r-edges from A-nodes to B-nodes are transformed into s-edges from A-nodes to B-nodes; then each s-edge from an A-node to a B-node is replaced by a r-edge in the opposite direction from the B-node to the A-node. The program thus is made up of two rules: (1) *rename* which locally renames a r-edge between an A-node a and a B-node b, so that $a\ r\ b$ turns into $a\ s\ b$; (2) *reverse* which locally replaces a s-edge between an A-node a and a B-node b by a r-edge between b and a so that $a\ s\ b$ turns into $b\ r\ a$. The main of the program calls, in an iterative way, first the rule *rename* then *reverse*.

The question is how to prove that the given program produces the expected states of the graph specified by *TFormulae*. This verification problem will be discussed in the next section.

3   Small-t$\mathcal{ALC}$ Program Verification

We are interested in verifying the correctness of transformation programs, i.e. checking whether a transformation behaves the way it is expected to and produces what it should. Therefore, besides verifying the correctness of each rule, we need to verify that the sequence of rules in the main program is also correct.

3.1   Motivating Example

For instance, consider the program in Fig. 2. To prove that the transformation is correct, the following points must be verified:

1. The correctness of the rules *rename* and *reverse* with respect to their *ABox* pre- and postconditions,
2. The correctness of applying iteratively the two rules *rename* and *reverse* with respect to the *TBox* assert clauses.

The second point necessitates examining global modifications in the host graph. The properties to be verified in (2) are global because they concern a set of nodes of type A or of type B thus they cannot be expressed with *AFormulae* but by *TFormulae*. We can specify the transformation program of Fig. 2 as follows: if in the source graph there are r-edges connecting A-nodes to B-nodes and there is not s-edges outgoing from A-nodes, then after applying iteratively the rules *rename* and *reverse*, there are r-edges connecting B-nodes to A-nodes and there is not r-edges nor s-edges outgoing from A-nodes.

More precisely, according to DL definitions [9], the *TFormula* $(\exists\ r^{-1}\ A) \subseteq B\ \wedge\ (\exists\ s^{-1}\ A) = \bot$ asserts that before transformation B-nodes subsume the target nodes of the r-edges outgoing from A-nodes and that the set of target nodes of the s-edges outgoing from A-nodes is empty. From this assumption, we verify after transformation the *TFormula* $(\exists\ r^{-1}\ A) = \bot\ \wedge\ (\exists\ r^{-1}\ B) \subseteq A$ which expresses now that the set of target nodes of the r-edges outgoing from A-nodes is empty and that A-nodes subsume the set of target nodes of the r-edges outgoing from B-nodes. After the transformation, $(\exists\ s^{-1}\ A) = \bot$ stays as an invariant to express the temporary use of s-edges which are created in the rule *rename* are deleted in the rule *reverse*.

This paper focuses on reasoning about global properties on concepts and roles, i.e. properties of the graph as a whole as in (2), that are impacted by application of a set of rules, one or many times. For this purpose, we provide reasoning capabilities not only at rule-level using *AFormulae* but also at program-level using *TFormulae*. Proving the correctness of a program entails verifying that both the source graph (an interpretation) and the target graph (another interpretation) are models of the *TBox* and *ABox*. The next sections presents our solution to verify a rule at *ABox* level and to verify a program at *TBox* level.

3.2   Rule Verification Using *ABox* Layer

Within a rule, Small-t$\mathcal{ALC}$ uses *AFormulae* to specify graph elements manipulated by the rule's code in the pre- and postconditions. Therefore, only named graph nodes and edges in the current matched graph are concerned. In other words, a rule-level verification allows reasoning only about the local effect when applying once a rule on a matched graph.

Adopting Hoare-like calculus, a prover was developed [7,10] to prove that a Small-t$\mathcal{ALC}$ rule $\{P\}S\{Q\}$ is correct. This verification process is based on an *ABox* static analysis performed in a backward mode in order to compute the weakest precondition (wp) [13]. Each rule statement s of S is assigned to a predicate transformer yielding an $\mathcal{ALCQI}$ formula $wp(s, Q)$ assuming the postcondition Q. The correctness of the code S of a rule with respect to Q is established by proving that the given precondition P implies the weakest precondition.

3.3   Program Verification Using *TBox* Layer

As stated in Sect. 2.3, *TFormulae* are implicitly updated by rules statements that explicitly add and delete individuals and pairs of individuals respectively into and from concepts interpretations. Reasoning about graph global properties when executing a sequence of *ABox* rules turns into studying the effects of *ABox* statements on the *TBox* properties. This results in verifying *TFormulae* of the transformation program in order to check if the graph is correctly transformed as expected.

Using *TFormulae*, we consider an abstract graph that is a superset of the concrete Small-t$\mathcal{ALC}$ graph: properties on nodes are ignored and only properties about the sets of nodes and the sets of source and target nodes of roles are taken into account. Considering such global properties results in losing certain information regarding *AFormulae*. For example, we can not know concretely each pair of connected nodes given the property "all r-edges outgoing from A-nodes go towards B-nodes" i.e. $(\forall\ r^{-1}\ A) \subseteq B$. This abstraction idea and its formalization is called the theory of abstract interpretation [14].

The main question in this paper is how to infer the *TBox* properties on abstract graphs thus allow verifying a program consisting of a sequence of *ABox* rules, not only at rule level as done in our previous work. In the next section, we present in detail our solution for this question.

4   Static Analysis by Abstract Interpretation

In order to verify the global state of a graph before and after rule applications, we study the impact of *ABox* Small-t$\mathcal{ALC}$ statements on a given *TFormula* representing *TBox* properties. To do so, we analyze the effect of adding (deleting) an element to (from) a set on the set equality and inclusion relationships.

4.1   Interpretation of Small-t$\mathcal{ALC}$ Statements

The aim of our proposed static analysis is to infer a *post-TFormula* on the basis of a given *pre-TFormula* considered as a rule's assumption by interpreting the rules statements in a forward chaining. The inference of a such *TFormula* is done by studying the effect of *add* and *delete* statements on each *TFact* in the *pre-TFormula* considering the statement's precondition as hypothesis. For instance, given the *TFact* $C = D$ in a *pre-TFormula*, adding an individual i to C through the instruction $add(i : C)$ may affect the validity of $C = D$. If i is already an element of C, according to set theory, $add(i : C)$ has no effect on the set C. Consequently, $C = D$ remains valid. However, if i does not belong to C, $add(i : C)$ will add one additional element to C, thus C becomes $C \cup \{i\}$. Consequently, $C = D$ turns into $C \supseteq D$. The *AFact* $i : C$ can be checked in the precondition of the statement $add(i : C)$.

To clarify the static analysis process, consider the inference of a *post-TFormula* after the call of the rule *rename* with respect to the *pre-TFormula* $(\exists\ r^{-1}\ A) \subseteq B\ \wedge\ (\exists\ s^{-1}\ A) = \bot$ given in the main of Fig. 2. As illustrated in Fig. 3, the pre- and postconditions of the statements are specified by computing the strongest postcondition (*sp*) of the statement from its precondition. The *sp* of a statement expresses most accurately the evolution of the graph being transformed at the *ABox* level. Taking into account these *ABox* effects on individual nodes and edges, we want to determine the most precise evolution, at *TBox* level, of the concepts and edges containing these individuals. The inference of a *post-TFormula* after each statement is done by studying the effect of the statement on the *TFormula* while taking into account the properties of the manipulated nodes identified in the *ABox* precondition of the statement. In this example, the statement *delete(a r b)* that removes the *r*-edge between the nodes a and b does not affect any *TFact* of the *pre-TFormula*. In fact, the deletion of the pair (b, a) from the set r^{-1}, knowing that $a : A$ and $b : B$, holds the validity of the inclusion $(\exists\, r^{-1}\, A) \subseteq B$ and does not concern the *TFact* $(\exists\, s^{-1}\, A) = \bot$ which remains valid. However, adding an *s*-edge between the nodes a and b, knowing that $a : A$ and $b : B$ from the statement's precondition, transforms the *TFact* $(\exists\, s^{-1}\, A) = \bot$ into $(\exists\, s^{-1}\, A) \subseteq B$ in the *post-TFormula*.

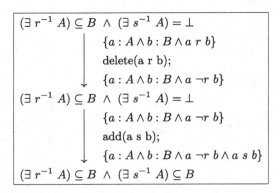

Fig. 3. Inference of a *TFormula* after each statement of the rule *rename*

Table 2 summarizes the effect of the statement *add(i : C)* on both equality and inclusion relationships between concepts. The second column presents the *pre-TFact*; the *AFacts* considered as hypothesis for the interpretation are shown in the third column. The fourth column provides the inferred *TFact* obtained by the interpretation, so called *post-TFact*. In cases where the *pre-TFact* is confirmed as being not valid yet the effect of the statement on the *pre-TFact* can not be deduced, that *TFact* will be deleted from the *TFormula*. This case is marked in the table by X. For instance, given the *pre-TFact* $C \subseteq D$, adding an instance i to the concept C with i not declared of concept C nor D may make the inclusion not valid. No more informations can be deduced to infer a *post-TFact* so it is deleted from the final *TFormula* so-called *post-TFormula*. Due to the limited number of pages allowed, the tables referring to the others atomic statements as well as the supporting tools are not presented here but available for download[1].

The *select* statement has no effect on a *TFormula* as it is an assignment of nodes variables. Whereas *if condition then s1 else s2* is interpreted by transforming the *pre-TFormula* regarding the sequence *s1* on the one hand, and *s2* on the other. The result is the disjunction of both of the resulting *post-TFormulae*.

The body of the *while* loop is interpreted once, as well as the body of a rule that is called in an iterative way in the main. In fact, whether the interpretation

---

[1] https://www.irit.fr/~Martin.Strecker/CLIMT/Software/smalltalc.html.

Table 2. Interpretation of the statement $add(i : C)$

Statement	pre-TFact	AFact	post-TFact
$add(i : C)$	$C = \bot$	-	$\neg(C = \bot)$
	$\neg(C = \top)$	$i : C$	$\neg(C = \top)$
		else	X
	$C = D$	$i : C$	$C = D$
		else	$D \subseteq C$
	$C \subseteq D$	$i : C \vee i : D$	$C \subseteq D$
		else	X
	$C \cup D = \bot$	-	$\neg(C \cup D = \bot)$
	$\neg(C \cup D = \top)$	$i : C \vee i : D$	$\neg(C \cup D = \top)$
		else	X
	$C \cup D = E$	$i : C \vee i : D \vee i : E$	$C \cup E = E$
		else	$E \subseteq C \cup D$
	$C \cup D \subseteq E$	$i : C \vee i : D \vee i : E$	$C \cup D \subseteq E$
		else	X
	$C \cap D = \bot$	$i : C \vee i : \neg D$	$C \cap D = \bot$
		else	$\neg(C \cap D = \bot)$
	$\neg(C \cap D = \top)$	$i : C \vee i : \neg D$	$\neg(C \cap D = \top)$
		else	X
	$C \cap D = E$	$i : C \vee i : E \vee i : \neg D$	$C \cap D = E$
		else	$E \subseteq C \cap D$
	$C \cap D \subseteq E$	$i : C \vee i : E \vee i : \neg D$	$C \cap D \subseteq E$
		else	X
	$(\exists\, r\, C) = D$	-	$D \subseteq (\exists\, r\, C)$
	$(\exists\, r\, C) \subseteq D$	-	X
	$(\exists\, r^{-1}\, C) = D$	$i : (= 0\, r\, \neg D)$	$(\exists\, r^{-1}\, C) = D$
		else	$D \subseteq (\exists\, r^{-1}\, C)$
	$(\exists\, r^{-1}\, C) \subseteq D$	$i : (= 0\, r\, \neg D)$	$(\exists\, r^{-1}\, C) \subseteq D$
		else	X

of the same sequence of statements is done one or several times, the resulting $TFormula$ remains the same as Small-t$\mathcal{ALC}$ statements are limited to adding and deleting elements to and from sets as already mentioned. For instance, consider a $TFact$ $C = D$ and a statement that adds repeatedly a selected instance d to the concept D. By interpreting the statement for the first time, $C = D$ turns into $C \subseteq D$. Adding other elements d to the concept D maintains the validity of the $TFact$ $C \subseteq D$. Consequently, the traditional widening operator of the abstract interpretation, which guarantees termination when applied to increasing sequences [14], is simpler in our context than in programs employing non symbolic operations.

4.2   Soundness of the Static Analysis

Deriving a $TFormula$ for Small-t$\mathcal{ALC}$ programs does not guarantee that our static verification calculus is sound. Given a correctness formula $\vdash \{P\}S\{Q\}$, we need to show that the proposition $\models \{P\}S\{Q\}$ about the semantics of the correctness formula holds. This entails to consider $\models \{P\}S\{Q\}$ as a new judgment based on state updates meaning that the program S when invoked in the state σ will terminate in the state τ. We denote $S(\sigma, \tau)$ this relation and define $\models \{P\}S\{Q\}$ as $\forall \sigma.P(\sigma) \Rightarrow (\exists \tau.S(\sigma, \tau) \wedge Q(\tau))$. Proof is done on the derivation of Hoare correctness formulae considering Small-t$\mathcal{ALC}$ operational semantics.

Let us consider $S = add(i : C)$, inspired from the assignment statement $V := E$ in imperative languages for which $sp(V := E, P) = \exists V'.P[V'\backslash V] \wedge (V = E[V'\backslash V])$, we compute $sp(add(i : C), P)$ as a substitution: $sp(add(i : C), P) = \exists C'.P[C'\backslash C] \wedge (C' + i\backslash C)$. If the formula $sp(add(i : C), P) \Rightarrow Q$ is valid, then for all source graphs G verifying $P(\sigma)$ we conclude $Q(\sigma')$ for target graphs G' where σ' denotes the state σ updated by the action add.

In the following, we prove two inferences about adding a node i to a concept C. The first one is basic and corresponds to the first line of Table 2. Suppose a state $\sigma = (C = \bot)$. Then $sp(add(i : C), C = \bot) = \exists C'.(C = \bot)[C'\backslash C] \wedge (C' + i\backslash C)$, that is $C' = \bot \wedge C = C' + i$ which implies $\sigma' = \neg(C = \bot)$. As $P(\sigma)$ is true for G, we have $Q(\sigma')$ for G'. Thus, $P(\sigma) \Rightarrow S(\sigma, \sigma') \wedge Q(\sigma')$. This case is quite straightforward because it does not presuppose any $AFact$ for P.

The second one assumes the precondition $C = D$, as indicated by the third line of Table 2. We aim at knowing when this relation of subsumption $TBox$ between the concepts C and D is also a postcondition of the substitution $[C' + i\backslash C]$ related to $add(i : C)$. The outcome of this question depends on whether the individual i belongs to concept C. If $\sigma = (C = D \wedge i \in C)$, we can conclude that $\sigma' = (C = D)$, otherwise, $\sigma = (C = D \wedge i \notin C)$ is transformed into $\sigma' = (C \supseteq D)$. In the first case, we have $sp(add(i : C), C = D \wedge i \in C) = \exists C'.(C = D \wedge i \in C)[C'\backslash C] \wedge (C' + i\backslash C)$, i.e. $C' = D \wedge i \in C' \wedge C = C' + i$ which implies $C = D$, because $i \in C' \wedge C = C' + i \Rightarrow C = C'$. On the other hand, when $i \notin C$, $sp(add(i : C), C = D \wedge i \notin C) = C' = D \wedge i \notin C' \wedge C = C' + i$ which implies $C \supseteq D$, because $i \notin C' \wedge C = C' + i \Rightarrow C \supseteq C'$. As previously, and in both cases, as $P(\sigma)$ is true for G, we have $Q(\sigma')$ for G'. Thus, $P(\sigma) \Rightarrow S(\sigma, \sigma') \wedge Q(\sigma')$.

We can prove the other lines of Table 2 similarly, considering the $ABox$ substitutions of the language and the $TFormulae$ involved.

5   Relation Between the ABox/TBox Verifications

The purposes of $ABox$ and $TBox$ verifications differ. $TBox$ verification aims to verify concepts inclusion relationships (universal assertions), whereas $ABox$ verification is more about fact-checking and instance-checking (membership assertions). In terms of program verification, they are complementary. However, these two components are undoubtedly dependent.

5.1   Dependence Between the ABox/TBox Verifications

Inferring $TFormulae$ does not consider only rules statements, but takes into account rule specifications on instances properties too. Therefore, weakening $AFormulae$ has a direct effect on the process of inferring $TFormulae$. In case where instances properties are not revealed in the precondition, some properties on sets may not be proven to be valid and so are discarded from the *post-TFormula*.

For instance, consider the program of Fig. 4 consisting of the rule *replace* which replaces a r-edge between two nodes a and b with s. The precondition of the rule asserts that b is a B-node, however, it does not inform about the concept of a. Note that this rule is proven to be correct by the Small-t$\mathcal{ALC}$ prover.

```
rule replace {
    pre: (a r b) ∧ (b : B)
    // strengthened pre: (a r b) ∧ (b : B) ∧ (a : A)
        delete(a r b);
        add(a s b);
    post: (a s b) ∧ (b : B)
}
main {
    assert: (∃ s⁻¹ A) = ⊥
        replace;
    assert: (∃ s⁻¹ A) ⊆ B
}
```

Fig. 4. Example of inconsistent $TFormulae$

Consider the $TFact$ $(\exists\ s^{-1}\ A) = \bot$ before the rule call expressing that there is no s-edges outgoing from A-nodes. Aiming for verifying after the rule call that edges outgoing from A-nodes are going towards B-nodes i.e. $(\exists\ s^{-1}\ A) \subseteq B$, the static analyzer studies the effect of the rule *replace*. So it interprets firstly the statement $delete(a\ r\ b)$ which does not affect the validity of $(\exists\ s^{-1}\ A) = \bot$, and secondly the statement $add(a\ s\ b)$ which certainly does because the given $TFact$ concerns the added s-edge. In this case, the static analyzer shows that the $TFact$ $(\exists\ s^{-1}\ A) = \bot$ is unsatisfiable, but does not infer any other fact since the concept of a is unknown (corresponding to a case X in the interpretation table of the statement $add(i\ r\ j)$). Hence, the given $TFact$ $(\exists\ s^{-1}\ A) \subseteq B$ is supposed inconsistent with $ABox$ assertions of the rule *replace*.

Now suppose that the developer asserts as well in the precondition of the rule that a is an A-node as shown in bold on Fig. 4. In this case, the static analyzer would deduce that $(\exists\ s^{-1}\ A) \subseteq B$. We can conclude that the more strengthened $AFormulae$ are, the more the diagnostic of $TFormulae$ gets refined.

5.2   Complementarity Between the ABox/TBox Verifications

Verification of a rule's triple using Hoare logic guarantees a correct transformation of the manipulated nodes. At a more abstract level, verification of the $TBox$ checks the effect of the rules on the graph as a whole. These two verification levels are complementary: each level verifies properties that can not be expressed by the other one.

Let us reconsider the program in Fig. 2 consisting in reversing r-edges outgoing from an A-node a towards B-nodes. Suppose now that the developer makes an error in the rule $reverse$ by writing the statement $add(a\ r\ b)$ instead of $add(b\ r\ a)$ as shown in bold in Fig. 5. In this case, the rule renames each s-edge to r-edge without reversing it. Consequently, applying the sequence $\{rename!, reverse!\}$ on a graph will produce a target graph identical to the source graph.

```
rule reverse {
    pre: (a : A) ∧ (b : B) ∧ (a s b)
        delete(a s b);
        add(a r b);
        // correct statement: add(b r a);
    post: (a : A) ∧ (b : B) ∧ (a ¬s b)
}
```

Fig. 5. Incorrect rule $reverse$

The rule $reverse$ is proven as a correct Hoare-triple by the Small-t$\mathcal{ALC}$ prover, i.e. the rule's code ensures the postcondition with the given precondition. This happens because the rule's postcondition is weak: it checks only the concept of a and the nonexistence of s-edges outgoing from a. However, exploiting the Small-t$\mathcal{ALC}$ $TBox$ static analyzer to verify the $post$-$TFormula$ given in the $main$ program, we notice that the $TFact$ $(\exists\ r^{-1}\ A) = \bot$, which expresses that there is not a r-edges outgoing from A-nodes, is unsatisfiable.

Warned by the result of the $TBox$ verification, the developer strengthens the postcondition of the rule $reverse$ with the $AFact$ $a : (= 0\ r\ B)$ to check that no r-edge is outgoing from a. Now the prover fails to verify the rule with the modified postcondition. The developer then realizes that b must be connected by r to a.

Since one is allowed to write weaken specifications of a code while maintaining the validity of a rule's triple at the $ABox$ level, a given postcondition may not reveal all the properties of the transformed instances to be verified yet yield to a correct triple. In those cases, using $TBox$ verification with $TFormulae$ can identify an abnormal effect on the graph.

On the other hand, verifying exclusively that the given $TFormulae$ are consistent with the global graph does not attest actually that rules triples are written correct since the *TBox* verification infer *TFormulae* from rules supposed correct. Hence, it is necessary as well to prove rules triples using Hoare logic by writing complete specifications to get tangible results.

Ultimately, each of the *ABox* and *TBox* verifications has different level of verification and so are complement. *ABox* checks whether instances that are manipulated in a rule are locally transformed. *TBox* checks the effect of instances transformation on the abstract graph. Hence, errors that are not identified by one level, can be identified by the other.

6   Verifying Monadic Second-Order Properties

Verifying rules using the Hoare logic with *ABox* assertions on individuals is limited for checking local properties of the graph. With quantification over sets, *TBox* assertions can express global properties of graphs and can be exploited to verify some monadic second-order (MSO) properties [6].

For instance, consider the problem of verifying that a graph is bipartite i.e. a graph that is colored in two colors e.g. A and B, and in which every edge connects a node of A to one of B. Figure 6 shows the Small-t$\mathcal{ALC}$ rule *grow* that allows connecting, with a r-edge, two nodes belonging to two different concepts. The bipartiteness property can be expressed in the Small-t$\mathcal{ALC}$ *TFormulae* by two *TFacts*: $(\exists\, r\, A) \cap A = \bot$ to verify that the set of source nodes of the r-edges going towards A and A are disjoint, and $(\exists\, r\, B) \cap B = \bot$ to verify that the set of source nodes of the r-edges going towards B and B are disjoint. To close off the possibility to add a r-edge outgoing from nodes belonging to other concepts than A or B, closure axioms are necessary: $A \cup B = \top \wedge A \cap B = \bot$ i.e. all the graph's nodes are exclusively of concept A or of concept B. This *TBox* invariant expressing a global property of the graph can be checked before and after calling iteratively the rule *grow*.

```
rule grow {
  pre: x : (= 0 r A ∪ B)
    if(x : A) then
      select y with y : B;
    else
      select y with y : A;
    add(x r y);
  post: x : (= 1 r A ∪ B)
}
main {
  assert: (∃ r A)∩A = ⊥ ∧ (∃ r B)∩B = ⊥∧ A∪B = ⊤∧A∩B = ⊥
  grow!;
  assert: (∃ r A)∩A = ⊥ ∧ (∃ r B)∩B = ⊥∧ A∪B = ⊤∧A∩B = ⊥
}
```

Fig. 6. Small-t$\mathcal{ALC}$ program making up a bipartite graph

Our *TBox* abstraction level neglects the source and target nodes of an edge. Hence, our current work is not able to express directly MSO properties related to connectivity of a graph. We envisage increasing the expressiveness of *TBox* formulae by choosing a richer description logic, notably which offer role constructors and role connectors such as inclusion and transitivity.

7 Related Work

In the theory of algebraic graph transformations, Habel and Pennemann [1] defined nested application conditions to describe graph properties. However these first-order tailored logic formulae need to be derived into specific inference rules in order to provide a specific theorem-proving that suits them. This approach has been adopted by the graph transformation language GP [15] which provides a Hoare-like calculus. Nested conditions of GP have been recently extended to MSO properties on graphs by introducing new quantifiers for set variables of nodes and edges and having morphisms with constraints about set membership [5].

The algebraic approach has also given rise to the dedicated logic for graph properties, called Graph Pattern Logic [16] and Navigational Logic [17], which consider that a graph pattern P is just an object in the category of graphs. Thereby, a global property for a graph G can be reduced to identifying a morphism from P to G. The authors have invested patterns dedicated to graph paths between nodes. We share with them the idea that reasoning mechanisms are supported by the underlying logic.

The static satisfiability of a DL knowledge base updated by a finite sequence of insertions and deletions performed on concepts and roles has been studied by Calvanese et al. [18]. The authors introduce a simple imperative language with the basic actions $A \oplus C$ and $A \ominus C$ on an interpretation $\mathcal{I}$ for concepts A and C. $A \oplus C$ stands for the addition of the content of $C^{\mathcal{I}}$ to $A^{\mathcal{I}}$ and $A \ominus C$ represents the removal of $C^{\mathcal{I}}$ from $A^{\mathcal{I}}$.

In order to capture the action effects on a DL knowledge base $\mathcal{K}$, a transformation $TR(\mathcal{K})$ associated to each action has been defined. This transformation on a finite interpretation domain enables to reduce static verification to finite satisfiability of $\mathcal{K}$: $TR(\mathcal{K})$ is $\mathcal{K}$-preserving if there exists a model when applying $TR(\mathcal{K})$ on interpretations. Transformations allow to modify labels of sets of nodes instead of individuals. Constraints on interpretations coding graph-structured data are expressed by specific $\mathcal{ALCHOIQ}$br DL formulae, including nominals ($\mathcal{O}$) which enables modifying single node labeling.

Dynamic logics [19] are well suited for dealing about properties of evolving data. J. H. Brenas et al. [20] investigate such logics for graph transformations and define $\mathcal{C2PDLS}$, a combination of both combinatory and converse propositional dynamic logics, augmented by substitutions. The main idea is to split the nodes of the considered graphs into two sets: one contains the nodes before substitutions take place; the other stores nodes that will be created by future transformations and those that have been deleted in the past. This separation allows some reasoning on reachability properties considering named nodes.

In our Small-t$\mathcal{ALC}$ context, $ABox$ updates do not represent changes or refinements in the conceptualization of $TBox$ axioms. We allow adding and deleting individuals and roles in an imperative style with extensional $ABox$ rules, while provide a mechanism to infer intentional $TBox$ knowledge which is consistent with $ABox$ changes. From the user point of view, we share the same desired effect called projection in action-oriented paradigm, i.e. knowing whether an assertion that one wants to make true really holds after executing a rule [21].

8   Conclusion and Future Work

Our logic-based graph transformation language Small-t$\mathcal{ALC}$ allows to reason on graph transformations and verify local and global properties of graphs by exploiting $ABox$ and $TBox$ levels of description logic respectively. The properties of nodes manipulated in each rule are expressed in $ABox$ pre- and postconditions so that a Hoare-like calculus can be realized to verify the correctness of a rule. Besides this $ABox$ verification, we presented an approach based on a static analysis aiming to deduce implicit $TBox$ assertions about concepts from explicit $ABox$ assertions and valid $TBox$ premises. Our $TBox$ verification process determines whether the given $ABox$ and $TBox$ assertions are consistent. A formal proof sketch of our static algorithm has been addressed.

We showed that using $TFormulae$, some monadic second-order properties can be verified. It would be interesting as future work to improve the expressiveness of our $TFormulae$ in such a way that more global properties can be verified e.g. considering the cardinality restrictions and roles constructors.

Other dialects and in particular DL $\mathcal{ALCQIO}$ with nominals $\mathcal{O}$ which allows the description of concepts by the enumeration of named individuals can be considered as well. The key is to work out how we can increase the expressivity of Small-t$\mathcal{ALC}$ programs in order to be able to prove more interesting specifications. We also investigate Small-t$\mathcal{ALC}$ functionalities to manage explicit and inalterable $TBox$ axioms now given by the end-user.

References

1. Habel, A., Pennemann, K.-H.: Correctness of high-level transformation systems relative to nested conditions. Math. Struct. Comput. Sci. **19**(2), 245–296 (2009). https://doi.org/10.1017/S0960129508007202
2. Rensink, A.: Representing first-order logic using graphs. In: Ehrig, H., Engels, G., Parisi-Presicce, F., Rozenberg, G. (eds.) ICGT 2004. LNCS, vol. 3256, pp. 319–335. Springer, Heidelberg (2004). https://doi.org/10.1007/978-3-540-30203-2_23
3. Orejas, F., Ehrig, H., Prange, U.: A logic of graph constraints. In: Fiadeiro, J.L., Inverardi, P. (eds.) FASE 2008. LNCS, vol. 4961, pp. 179–198. Springer, Heidelberg (2008). https://doi.org/10.1007/978-3-540-78743-3_14
4. Lambers, L., Orejas, F.: Tableau-based reasoning for graph properties. In: Giese, H., König, B. (eds.) Graph Transformation, pp. 17–32. Springer, Cham (2014)
5. Poskitt, C.M., Plump, D.: Verifying monadic second-order properties of graph programs. In: Giese, H., König, B. (eds.) Graph Transformation, pp. 33–48. Springer, Cham (2014)

6. Courcelle, B.: The expression of graph properties and graph transformations in monadic second-order logic. In: Handbook of Graph Grammars and Computing by Graph Transformations, Volume 1: Foundations, pp. 313–400 (1997)
7. Strecker, M.: Modeling and verifying graph transformations in proof assistants. Electron. Notes Theoret. Comput. Sci. **203**(1), 135–148 (2008)
8. Baklanova, N., et al.: Coding, executing and verifying graph transformations with small-t$\mathcal{ALCQ}$e. In: 7th International Workshop on Graph Computation Models(GCM) (2016). http://gcm2016.inf.uni-due.de/
9. Baader, F., Calvanese, D., McGuinness, D.L., Nardi, D., Patel-Schneider, P.F. (eds.): The Description Logic Handbook: Theory, Implementation, and Applications. Cambridge University Press, New York (2003)
10. Makhlouf, A., Percebois, C., Tran, H.N.: An auto-active approach to develop correct logic-based graph transformations. Int. J. Adv. Softw. **11**(1,2), 147–158 (2018) http://oatao.univ-toulouse.fr/22689/
11. Sattler, U.: Reasoning in description logics: basics, extensions, and relatives. In: Antoniou, G., et al. (eds.) Reasoning Web 2007. LNCS, vol. 4636, pp. 154–182. Springer, Heidelberg (2007). https://doi.org/10.1007/978-3-540-74615-7_2
12. De Giacomo, G., Lenzerini, M., Poggi, A., Rosati, R.: On instance-level update and erasure in description logic ontologies. J. Logic Comput. **19**(5), 745–770 (2009)
13. Dijkstra, E.W., Scholten, C.S.: Predicate Calculus and Program Semantics. Springer, New York (1990). https://doi.org/10.1007/978-1-4612-3228-5
14. Cousot, P.: Abstract interpretation based formal methods and future challenges. In: Wilhelm, R. (ed.) Informatics. LNCS, vol. 2000, pp. 138–156. Springer, Heidelberg (2001). https://doi.org/10.1007/3-540-44577-3_10
15. Poskitt, C.M., Plump, D.: Hoare-style verification of graph programs. Fundam. Inform. **118**, 135–175 (2012)
16. Navarro, M., Pino, E., Orejas, F., Lambers, L.: A logic of graph conditions extended with paths. In: Pre-proceedings 7th International Workshop on Graph Computation Models (2016). http://gcm2016.inf.uni-due.de/pre-proceedings.html
17. Lambers, L., Navarro, M., Orejas, F., Pino, E.: Towards a navigational logic for graphical structures. In: Heckel, R., Taentzer, G. (eds.) Graph Transformation, Specifications, and Nets. LNCS, vol. 10800, pp. 124–141. Springer, Cham (2018). https://doi.org/10.1007/978-3-319-75396-6_7
18. Ahmetaj, S., Calvanese, D., Ortiz, M., Simkus, M.: Managing change in graph-structured data using description logics. ACM Trans. Comput. Logic **18**(4), 27:1–27:35 (2017). https://doi.org/10.1145/3143803
19. Harel, D., Kozen, D., Tiuryn, J.: Dynamic Logic. In: Gabbay, D.M., Guenthner, F. (eds.) Handbook of Philosophical Logic. Handbook of Philosophical Logic, vol. 4, pp. 99–217. Springer, Dordrecht (2002). https://doi.org/10.1007/978-94-017-0456-4_2
20. Brenas, J.H., Echahed, R., Strecker, M.: C2PDLS: a combination of combinatory and converse PDL with substitutions. In: Gammarth, T., Mosbah, M., Rusinowitch, M. (eds.) 2017 the 8th International Symposium on Symbolic Computation in Software Science, SCSS 2017, 6–9 April 2017, pp. 29–41 (2017). https://easychair.org/publications/paper/dx4z
21. Liu, H., Lutz, C., Miličić, M., Wolter, F.: Reasoning about actions using description logics with general TBoxes. In: Fisher, M., van der Hoek, W., Konev, B., Lisitsa, A. (eds.) JELIA 2006. LNCS (LNAI), vol. 4160, pp. 266–279. Springer, Heidelberg (2006). https://doi.org/10.1007/11853886_23

Tools and Applications

Incremental (Unidirectional) Model Transformation with eMoflon::IBeX

Nils Weidmann[1(✉)], Anthony Anjorin[1], Patrick Robrecht[2], and Gergely Varró[2]

[1] Paderborn University, Paderborn, Germany
{nils.weidmann,anthony.anjorin}@uni-paderborn.de
[2] Paderborn, Germany

Abstract. Graph transformation is a mature formalism often used as a basis for model transformation tools. Although numerous graph transformation tools exist, very few explore the paradigm of reactive, event-driven programming via *incremental* graph transformation. As we believe reactive programming to be a promising application for graph transformation in both research and teaching, we have developed eMoflon::IBeX as a suitable environment for incremental unidirectional model transformation via graph transformation. With eMoflon::IBeX, we have realised a novel mix of complementary tool features that have proven to be useful and effective in predecessor tools. We discuss these features and present insights based on an empirical evaluation of eMoflon::IBeX.

Keywords: Graph transformation · Incremental pattern matching

1 Introduction and a Brief History

Graph transformation (GT) is a mature formalism often used as a formal underpinning for model transformation tools in the context of model-driven engineering. While numerous GT tools exist, many of which are still under active development, we have observed that very few explore the paradigm of *reactive*, event-driven programming via *incremental* GT [4]. Based on our work on implementing a novel Triple Graph Grammar (TGG) tool that leverages incremental graph pattern matching [11], we have come to regard reactive, event-driven programming to be a promising application for GT in both research and teaching, which has not yet received enough attention.

In this paper, therefore, we present *eMoflon::IBeX* as part of the eMoflon tool suite,[1] which has evolved out of a long line of predecessor GT tools. Figure 1 depicts the history of eMoflon, showing preceding and some related tools. Nodes represent tools, while edges indicate that one tool (successor) conceptually or/and technically evolved from another (predecessor). All edges ultimately leading to IBeX are labelled, indicating the primary reason for the evolution. Related tools are greyed out, while tools that are currently part of the eMoflon tool suite are highlighted with a light-blue background.

---

[1] www.emoflon.org.

© Springer Nature Switzerland AG 2019
E. Guerra and F. Orejas (Eds.): ICGT 2019, LNCS 11629, pp. 131–140, 2019.
https://doi.org/10.1007/978-3-030-23611-3_8

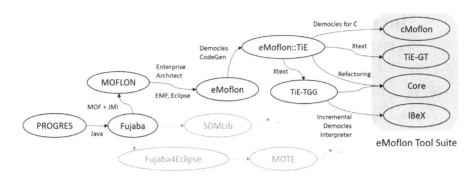

Fig. 1. History of eMoflon

Starting with PROGRES, one of the first tools for programmed GT, Fujaba was developed based on a mainstream GPL (Java). With the goal of implementing the full MOF[2] 2.0 and JMI[3] standard, MOFLON was developed as a plugin for Fujaba. With the success of Eclipse as an IDE platform, and EMF/Ecore as a *de facto* modelling standard, eMoflon was developed as a complete re-engineering of MOFLON. In addition, Enterprise Architect (EA)[4] was established as a visual front end for GT. For further details concerning the history of eMoflon, we refer to Anjorin et al. [1]. In its back end, eMoflon was still using the pattern matcher of Fujaba, which became increasingly challenging to evolve and maintain, partly because it was bootstrapped with a different tool chain. Based on Democles as a new pattern matcher [16], eMoflon::TiE was developed as a Democles-based version of eMoflon. In addition to providing a unified platform for both an interpretative and generative approach to model transformation, Democles was also designed to simplify exchanging all templates for code generation. This was exploited to establish cMoflon [10], a GT tool that generates embedded C code. While EA proved to be a scalable and relatively usable front end for eMoflon, it required a separate tool chain based on C# and Visual Studio. Combined with problems concerning licensing and cross-platform support, a decision was made to switch to Xtext[5] as an editor framework, and use PlantUML[6] for generated, read-only visualisations. This led to eMoflon::TiE-TGG for TGGs as a pilot project, and some time later, eMoflon::TiE-GT for GT. In this process, we extracted a common core component, eMoflon::Core, providing basic support for EMF code generation, and visualisation of EMF metamodels and models. For further details on this EA to Xtext migration, we refer to Yigitbas et al. [19].

Finally, driven by our requirements for TGGs [11], we developed eMoflon:: IBeX based on the incremental Democles interpreter [17]. eMoflon::IBeX realises

---

[2] The Meta Object Facility.
[3] Java Metadata Interface.
[4] www.sparxsystems.de.
[5] www.eclipse.org/Xtext/.
[6] http://plantuml.com/en/eclipse.

a novel mix of tool features that have proven to be effective over the years. In the following (Sect. 2), we discuss this mix of features and compare IBeX to related GT tools. We provide an architectural overview of IBeX in Sect. 3, with some details on specific tool features. The results of an empirical experiment conducted with 40 students are discussed in Sect. 4, while Sect. 5 concludes.

2   Motivation and Related Work

Table 1 depicts the six most important requirements that influenced the development of IBeX, and provides a comparison with five GT tools, including its predecessor TiE. Our choice of requirements was primarily driven by current research and teaching activities at TU Darmstadt and Paderborn University including model synchronisation, consistency checking and other model management tasks based on TGGs [2]. The comparison is not meant to be complete; there are at least double as many GT tools described in the literature – the chosen tools are close enough to IBeX to provide an interesting comparison and help to put IBeX in context to existing tools. Our comparison also does not imply that IBeX supports a superset of all features of the selected tools, for example, TiE is a code generator while IBeX is not, Henshin provides static analyses that IBeX does not, GRAPE supports schema-less GT while IBeX does not.

Table 1. Comparison of graph transformation tools

Tool, Version → Tool Feature ↓	EMorF [2.32.0]	GRAPE [0.4.2]	Henshin [0.1.1]	TiE [1.4.0]	VIATRA [2.1.1]	IBeX [1.0.0]
R1: Incrementality	✗	✗	✗	✗	✓	✓
R2: Full GT Paradigm	✓	✓	✓	✓	✗	✓
R3: TGG Integration	✓	✗	✗	✓	✗	✓
R4: GPL Integration	(✓)	✓	(✓)	✓	✓	✓
R5: Rule Modularity	✗	✗	✓	(✓)	✗	✓
R6: Textual + Visual	✗	✓	(✓)	✗	✗	✓

We require support for *incrementality* (R1) to enable a reactive, event-driven style of programming with GT. Our interest in exploring and supporting incrementality is shared by numerous approaches [5,7,14]. While there are tools and languages such as [13] that support reactive programming, to the best of our knowledge, (R1) is currently only fulfilled for GT tools by Viatra [15].

As we want to use and teach the GT approach, we require support for the *full GT paradigm* (R2), i.e., not only graph pattern matching, but also creation and deletion in form of GT rules. All GT tools in our comparison apart from Viatra provide this support; Viatra supports graph patterns with application conditions as rule preconditions, but chooses to provide a flexible DSL over Xtend for the "action" part of the rule, instead of the "green" (create) and "red" (delete) parts of a GT rule. We believe this sacrifices a substantial part of the simplicity and elegance of GT and makes it difficult to visualise and analyse entire rules meaningfully.

Due to our research and teaching focus on TGGs, we value a seamless integration of GT and TGGs (R3): users are able to mix and switch between GT and TGG, and TGG developers can reuse functionality of the GT layer. GT and TGG specifications should also use a consistent textual concrete syntax, visualisation, and project structure. Of all the tools in our comparison, only EMorF [9] and TiE [12] fulfil (R3).

A seamless integration with a mainstream GPL (R4) is crucial for practical applications, allowing users to easily mix and integrate GT rules into GPL code with support for code completion and type checking. While this is supported by all GT tools in our comparison, the GPL integration of EMorF and Henshin [6] is untyped and relies on rule names. This makes it difficult to check for errors at compile time.

Of similar importance for practical applications is dedicated support for modularity on the *level of GT rules* (R5). While Viatra supports advanced modularity concepts on the level of patterns, this does not cover the Xtend code for actions. Henshin provides variability-based reuse on the level of GT rules by using annotations. TiE supports rule refinement [8] only for TGG rules; This modularity feature is generalised in IBeX to uniformly cover patterns, GT rules, and TGG rules.

Finally, we are convinced that an Xtext-based textual editor combined with a read-only visualisation focused on the current position in the text editor (R6), allows for an efficient, effective editing experience that is also sustainable with respect to the cost of maintenance. From the tools in our comparison, this editing style is fully implemented only by GRAPE [18]. Although Henshin supports both a textual and visual concrete syntax, separate editors are used and have to be maintained separately.

3   Architectural Overview

eMoflon::IBeX is implemented as a set of Eclipse plugins and supports both incremental unidirectional model transformation with graph transformation, and bidirectional model transformation with TGGs. In this paper, however, we focus on the support for general GT. Figure 2 provides an architectural overview of IBeX: The TGG layer makes use of the GT layer, which consists of a front end and a back end component. The front end consists of an Xtext-based editor combined with a read-only visualisation using PlantUML. As input to the front

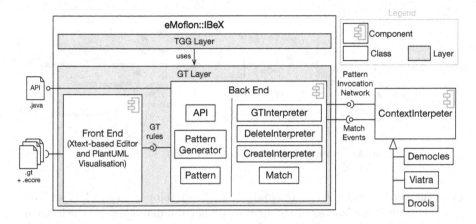

Fig. 2. Most important components and classes in eMoflon::IBeX

end, end-users provide `.ecore` files for all metamodels, and `.gt` containing graph transformation rules in a textual concrete syntax. We refer the interested reader to our handbook on unidirectional model transformation with IBeX [3] for screen shots of the visualisation and all details to the textual concrete syntax.

The front end produces GT rules (as EMF models) and passes them to the back end. Figure 2 depicts the most important interfaces and classes in the back end divided into compile time (to the left) and runtime (to the right).

At compile time, the back end uses a `PatternGenerator` to generate a set of separate `Patterns` from a GT rule. These patterns represent the context to be matched, elements to be deleted, and elements to be created. A typed `API` specially tailored for the set of GT rules is generated as Java code and produced as output for the end-user. This API wraps all calls to the `GTInterpreter` allowing for type safe access and rich compiler errors if rules are changed inconsistently.

At runtime, the `GTInterpreter` delegates the task of pattern matching to a `ContextInterpreter` as a separate component. A so-called *pattern invocation network* (an acyclic graph with patterns as nodes and invocations as edges) is passed, and events signalling new and invalid matches are expected. To use eMoflon::IBeX, an adapter for an incremental pattern matching engine is necessary. Currently, we mainly support Democles [16], but also have prototypes for Viatra and Drools[7]. The `GTInterpreter` collects all `Matches` and performs rule application by delegating deletion to a `DeleteInterpreter` and creation to a `CreateInterpreter`. While IBeX supplies default implementations for deletion and creation, these can be extended or replaced for special cases or optimisations.

Figure 3 depicts a communication diagram representing the GT rule application process at runtime. (1) The generated API serves as a factory for GT rules, providing methods for all non-abstract rules. (2) Rules can be used to subscribe for appearing or disappearing matches reported by the `GTInterpreter`. Rules

---

[7] www.drools.org.

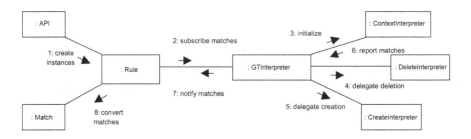

Fig. 3. Communication between API and GT interpreter

wrap the generic interpreter to avoid casting in developer code. (3) The interpreter initialises the `ContextInterpreter` for pattern matching, (4) the `Delete-Interpreter` for deletion, and (5) the `CreateInterpreter` for creation. When the monitored models are manipulated, (6) the `ContextInterpreter` produces and reports generic match events. (7) The `GTInterpreter` notifies the rule, which then (8) converts the generic matches to typed matches and provides them to the user via a series of methods such as `findAnyMatch` or `forEachMatch`, designed to work together with the standard Java stream API.

4   Evaluation

To evaluate IBeX and get feedback from users to further improve the tool, we conducted an empirical study with 40 students of an undergraduate, introductory course on model-based software development at Paderborn University. As part of the course, the students formed groups of 5–6 and used IBeX intensively for a semester while working on a small MBSE project involving DSL development.[8] We designed an online questionnaire[9] as a mix of quantitative multiple choice and qualitative open questions, to investigate the following research questions:

(Q1) *How do users perceive the editing experience provided by a combination of textual concrete syntax and coupled, read-only, partial visualisation?*

(Q2) *How do users judge the ease with which rules and patterns can be mixed with Java code and integrated in Java applications?*

(Q3) *How do users rate the relative importance of different language features?*

(Q4) *Do users appreciate our current documentation of IBeX as a set of handbooks realised as interactive Dropbox Paper[10] documents?.*

Figure 4 depicts an overview of the results from the quantitative part of the survey. All detailed results of the entire experiment are available online.[11] To

---

[8] http://bit.ly/2XT2fZ7.

[9] https://bit.ly/2VV8hGJ.

[10] With Dropbox Paper (https://www.dropbox.com/en_GB/paper), readers can communicate with authors via questions-and-answer threads directly integrated in the web-based document.

[11] https://bit.ly/2VUCrdc.

investigate our four questions, we formulated 23 multiple choice questions divided up into 5 categories. A scala of 1 to 5 was used for each question with 1 for "low" and 5 for "high". The first category *Prior Experience* was used to characterise our participants: programmers with sufficient experience with a modern object-oriented language, moderate prior experience with Eclipse, but with little to no prior experience with MDE, GT, or any visual language at all.

Regarding (Q1), our results indicate that many users find the textual concrete syntax acceptable, and even more appreciate the visualisation. While some users criticise the fact that the visualisation is read-only, our results show that it is probably not worth developing a visual editor, especially considering that most users are satisfied with the mix of a textual syntax and a coupled visualisation that adjusts dynamically to and focusses only on the current selection in the textual editor. By using the Xtext framework, our results show that it is possible to provide adequate validation errors and other usability features.

Concerning (Q2), our results indicate that while the expressiveness of the rule and pattern language is judged to be high enough, most students are uncertain if and how IBeX can be used in real-world applications. While this is probably due to some extent to a lack of experience (students of an undergraduate course), it still indicates that our current documentation tends to introduce the tool in isolation and should be improved to cover challenges involving an integration with other (UI, database, Web) frameworks for building realistic applications.

Regarding the integration of Java and GT code, being able to switch seamlessly between Java and GT files was judged to be acceptable but in need of improvement. The automatically generated JavaDoc for the API is appreciated by only a few users; most are neutral and apparently do not see any direct benefit.

Regarding (Q3), most students regard (positive and negative) application conditions and attribute conditions to be most important, followed by support for modularity (rule refinement), and complex application conditions (combination of conditions via conjunction (&&) and disjunction(||)). Many students are apparently unable to appreciate the potential of incrementality and reactive programming.

Finally, our handbook [3] (Q4) received mostly positive feedback, with many students preferring the example-driven, tutorial-like explanation to the complete, but reference-like appendix.

Threats to validity: Even though participation in the survey was anonymous and optional, we cannot completely exclude the fact that the students involved knew that their lecturer was one of the primary developers of the tool. This could have had a positive or negative effect as the students might have projected their (unrelated) satisfaction or frustration with the lecturer and the rest of the course on their answers.

Based on the relatively small sample of 40 students, we can only surmise general indications and suggestions for improvement of the tool. Generalising to other tools would require more data points and a more advanced experimental setup involving, e.g., a set of carefully prepared tasks and a control group.

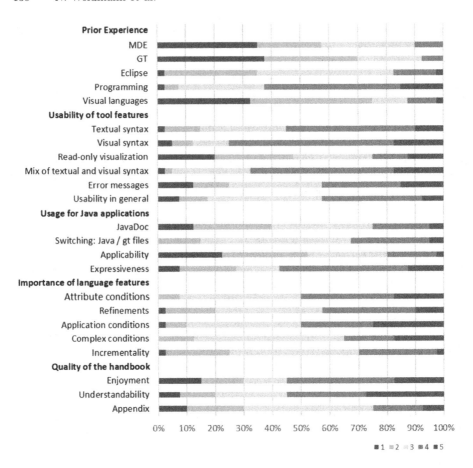

Fig. 4. Evaluation results

5 Conclusion and Future Work

We presented eMoflon::IBeX, a new GT tool with a special focus on supporting reactive programming via the incrementality of its underlying and exchangeable graph pattern matching engine. We discussed our most important requirements and compared IBeX with other related GT tools.

Based on the results of our empirical evaluation of IBeX, we plan to provide an additional handbook containing a series of applications and case studies using IBeX, chosen especially to showcase the advantages of reactive programming via the support for incrementality. We also plan to further improve the integration with Java, e.g., enabling a jump from API code back to the GT files containing the relevant rules and patterns. Concerning an empirical evaluation of GT tools, it would be interesting to evaluate different GT tools using, e.g., the System Usability Scale (SUS) questionnaire, which would enable a systematic comparison accross different tools. Finally, we are working on improving the scalability of the tool, which requires understanding how best to structure the generated pattern invocation networks passed to the incremental pattern matcher, taking the nature of the involved metamodels, the size of the models, and the size and connectivity of all patterns into account.

References

1. Anjorin, A., Lauder, M., Patzina, S., Schürr, A.: eMoflon: leveraging EMF and professional CASE tools. In: Informatik 2011, p. 281 (2011)
2. Anjorin, A., Leblebici, E., Schürr, A.: 20 years of triple graph grammars: a roadmap for future research. ECEASST **73** (2015)
3. Anjorin, A., Robrecht, P.: Unidirectional model transformation with eMoflon::IBeX (2018). https://bit.ly/2Hw1zDa
4. Bergmann, G., Ráth, I., Varró, G., Varró, D.: Change-driven model transformations. SoSyM **11**(3), 431–461 (2012)
5. Beyhl, T., Giese, H.: Incremental view maintenance for deductive graph databases using generalized discrimination networks. In: Heußner, A., Kissinger, A., Wijs, A. (eds.) GaM@ETAPS 2016. EPTCS, vol. 231, pp. 57–71 (2016)
6. Biermann, E., Ermel, C., Taentzer, G.: Formal foundation of consistent EMF model transformations by algebraic graph transformation. SoSyM **11**(2), 227–250 (2012)
7. Fritsche, L., Kosiol, J., Schürr, A., Taentzer, G.: Efficient model synchronization by automatically constructed repair processes. In: Hähnle, R., van der Aalst, W. (eds.) FASE 2019. LNCS, vol. 11424, pp. 116–133. Springer, Cham (2019). https://doi.org/10.1007/978-3-030-16722-6_7
8. Klar, F., Königs, A., Schürr, A.: Model transformation in the large. In: ESEC-FSE 2007, pp. 285–294. ACM, New York (2007)
9. Klassen, L., Wagner, R.: EMorF - a tool for model transformations. ECEASST **54** (2012)
10. Kluge, R., Stein, M., Giessing, D., Schürr, A., Mühlhäuser, M.: cMoflon: model-driven generation of embedded C code for wireless sensor networks. In: Anjorin, A., Espinoza, H. (eds.) ECMFA 2017. LNCS, vol. 10376, pp. 109–125. Springer, Cham (2017). https://doi.org/10.1007/978-3-319-61482-3_7
11. Leblebici, E., Anjorin, A., Fritsche, L., Varró, G., Schürr, A.: Leveraging incremental pattern matching techniques for model synchronisation. In: de Lara, J., Plump, D. (eds.) ICGT 2017. LNCS, vol. 10373, pp. 179–195. Springer, Cham (2017). https://doi.org/10.1007/978-3-319-61470-0_11
12. Leblebici, E., Anjorin, A., Schürr, A.: Developing eMoflon with eMoflon. In: Di Ruscio, D., Varró, D. (eds.) ICMT 2014. LNCS, vol. 8568, pp. 138–145. Springer, Cham (2014). https://doi.org/10.1007/978-3-319-08789-4_10
13. Perez, S.M., Tisi, M., Douence, R.: Reactive model transformation with ATL. Sci. Comput. Program. **136**, 1–16 (2017)
14. Schneider, S., Lambers, L., Orejas, F.: A logic-based incremental approach to graph repair. In: Hähnle, R., van der Aalst, W. (eds.) FASE 2019. LNCS, vol. 11424, pp. 151–167. Springer, Cham (2019). https://doi.org/10.1007/978-3-030-16722-6_9
15. Varró, D., Bergmann, G., Hegedüs, Á., Horváth, Á., Ráth, I., Ujhelyi, Z.: Road to a reactive and incremental model transformation platform: three generations of the VIATRA framework. SoSyM **15**(3), 609–629 (2016)

16. Varró, G., Anjorin, A., Schürr, A.: Unification of compiled and interpreter-based pattern matching techniques. In: Vallecillo, A., Tolvanen, J.-P., Kindler, E., Störrle, H., Kolovos, D. (eds.) ECMFA 2012. LNCS, vol. 7349, pp. 368–383. Springer, Heidelberg (2012). https://doi.org/10.1007/978-3-642-31491-9_28
17. Varró, G., Deckwerth, F.: A rete network construction algorithm for incremental pattern matching. In: Duddy, K., Kappel, G. (eds.) ICMT 2013. LNCS, vol. 7909, pp. 125–140. Springer, Heidelberg (2013). https://doi.org/10.1007/978-3-642-38883-5_13
18. Weber, J.H.: GRAPE – a graph rewriting and persistence engine. In: de Lara, J., Plump, D. (eds.) ICGT 2017. LNCS, vol. 10373, pp. 209–220. Springer, Cham (2017). https://doi.org/10.1007/978-3-319-61470-0_13
19. Yigitbas, E., Anjorin, A., Leblebici, E., Grieger, M.: Bidirectional method patterns for language editor migration. In: Pierantonio, A., Trujillo, S. (eds.) ECMFA 2018. LNCS, vol. 10890, pp. 97–114. Springer, Cham (2018). https://doi.org/10.1007/978-3-319-92997-2_7

Knowledge Representation and Update in Hierarchies of Graphs

Russ Harmer$^{(\boxtimes)}$ and Eugenia Oshurko

Univ. Lyon, EnsL, UCBL, CNRS, LIP, 69342 Lyon Cedex 07, France
{russell.harmer,ievgeniia.oshurko}@ens-lyon.fr

Abstract. A mathematical theory is presented for the representation of knowledge in the form of a directed acyclic hierarchy of objects in a category where all paths between any given pair of objects are required to be equal. The conditions under which knowledge update, in the form of the sesqui-pushout rewriting of an object in a hierarchy, can be propagated to the rest of the hierarchy, in order to maintain all required path equalities, are analysed: some rewrites must be propagated forwards, in the direction of the arrows, while others must be propagated backwards, against the direction of the arrows, and, depending on the precise form of the hierarchy, certain composability conditions may also be necessary.

Keywords: Knowledge representation · Graph rewriting · Graph databases

1 Introduction

We present a framework for knowledge representation (KR) based on hierarchies of objects from an appropriately structured category: a hierarchy is a directed acyclic graph (DAG) whose nodes are objects of the category and whose edges are arrows of the category such that all paths between each pair of objects are equal; we refer to this as the *commutativity* condition. The principal model of interest to us in this paper uses (simple) graphs and homomorphisms so that a hierarchy is a DAG whose nodes are themselves (simple) graphs. In this model, an edge of the DAG $h : G \to T$ asserts that the graph G is *typed* by T, i.e. T defines the kinds of nodes and kinds of edges (and attributes, if desired) that exist in G and h specifies, for each node and edge of G, which kind it is. As such, T can be viewed as a more abstract representation of knowledge of which G provides a more concrete instantiation.

We require certain structure on the category in order to be able to perform sesqui-pushout rewriting [4] to update an object in the hierarchy. However, such an update may invalidate some of the typing arrows of the hierarchy. The main contribution of this paper is to present a mathematical theory that guarantees the reconstruction of a valid hierarchy, after an arbitrary rewrite of an object, by appropriately *propagating* that rewrite to the other objects in the hierarchy.

E. Guerra and F. Orejas (Eds.): ICGT 2019, LNCS 11629, pp. 141–158, 2019.
https://doi.org/10.1007/978-3-030-23611-3_9

In the case where there are multiple paths between a given pair of objects of the hierarchy, this reconstruction depends on the satisfaction of a *composability* condition that guarantees that the propagated rewrites are compatible.

Motivating Use Cases

Modern database systems are increasingly migrating towards graph-based representations as a response to the growing wealth of data—from domains as varied as social or transport networks, the semantic web or biological interaction networks—that are most naturally expressed in those terms. However, unlike traditional relational DBs or earlier graph-based formats such as RDF, most graph DBs based on the richer model of property graphs [2,6] do not provide a native notion of *schema*. Our notion of hierarchy provides a mathematical framework for this. Indeed, an explicitly given schema graph to which a data, or instance, graph is homomorphic is the simplest non-trivial example of a hierarchy in our sense: the nodes of the schema specify the types of entites allowed in the system; its edges specify which edges between different types of nodes are allowed; and the attributes on its nodes and edges define the set of permitted attributes for nodes and edges. As such, the existence of a homomorphism from a data graph to a schema graph provides a proof of schema *validation*.

Our theory of propagation of rewriting in a hierarchy precisely captures the ways in which schema-aware DBs can be updated: a *descriptive* update occurs when the data is modified and the schema has to adjust accordingly; while a *prescriptive* update occurs when the schema is modified and the data needs to be adjusted. More precisely, if we *add* a node to the data graph and choose not to specify that its type already exists in the schema graph, in order to maintain the homomorphism from data to schema, we must propagate this operation to the schema graph to create a new node in the schema graph to type the new node of the data graph; similarly, if we *merge* two nodes of different types of the data graph, we must merge the corresponding typing nodes of the schema. Conversely, if we *delete* a node of the schema graph, we can only maintain the homomorphism by deleting all instances of that node in the data graph; and if we *clone* a node of the schema and choose not specify how to retype its instances in the data graph, those instances must be cloned in the data graph. In summary, *add* and *merge* updates propagate *forwards*, in the direction of the typing homomorphism, while *clone* and *delete* updates propagate *backwards*; and, as we will show, these observations remain true for general hierarchies.

Our theory thus provides a specification of how to enforce an *abstraction barrier* on a schema-less graph DB that provides the illusion of being schema-aware. Our Python library `ReGraph` implements this for the Neo4j graph DB by fixing an encoding of the data and schema graphs and the typing homomorphism within the single graph provided by Neo4j and translating any combination of clone, delete, add and merge operations into a corresponding query written in the Cypher language used by Neo4j [2]. More importantly, our theory also provides a specification of how to enforce the abstraction barrier corresponding to an *arbitrary* hierarchy—modulo the need to fix the encoding into Neo4j and the translation of update operations into Cypher. However, these two requirements are generic and can be derived systematically. As such, we provide the foundations for exploiting Neo4j (or similar graph DBs) as a platform for arbitrary user-defined KR systems.

The KAMI bio-curation system [8] has a richer 3-level hierarchy. At the root lies its *meta-model*, a fixed, hard-wired graph which defines the universe of discourse pertinent to the rule-based modelling of protein-protein interactions (PPIs) in cellular signalling: genes, regions of genes, binding and enzymatic actions, &c. The meta-model types an *action graph* which defines the particular collection of genes (and so on) of interest to a *corpus* of knowledge, e.g. a signalling pathway. The action graph types a *nugget graph*, containing many connected components, each representing the detailed conditions needed for a particular PPI to occur. In other words, an action graph summarizes the *anatomy* of a system while a nugget graph provides the *physiology* that determines how the system can behave.

In general, an update of the nugget graph refers to some anatomic features that already exist in the action graph and to others that must be added to maintain typing; this is performed automatically by forward propagation. It is important that propagation does not continue to the meta-model (which must remain unchanged); this is achieved by requiring that all new anatomic features specify how they are to be typed by the meta-model. This is an example of the notion of *controlled* forward propagation, discussed in Sect. 3, analogous to a descriptive DB update which actually preserves the current schema.

A knowledge corpus in KAMI can be contextualized, with respect to a choice of gene products, through an update of its action graph, giving rise to what we call a KAMI *model*; in the terminology of DBs, this is analogous to a *materialized view*—a contextualized copy of part of the original DB that can be manipulated independently. The effect of this update propagates backwards to the nugget graph. This propagation is not controlled—the cloning of a gene precisely gives rise to multiple gene products—unlike the case of *concept refinement* where the cloning of a schema node is accompanied by a specification of how to retype all instances of the original node in the data graph in terms of the refined schema. We discuss controlled backward propagation further in Sect. 4.

Neither of the above use cases require composability conditions, as discussed in Sect. 5, to guarantee valid reconstruction of a hierarchy after an update because they contain no (undirected) cycles—although this was actually exploited in an earlier version of KAMI and we anticipate that many other use cases will arise naturally.

Related Work

Slice categories provide many rich models of *typed* sesqui-pushout rewriting [4], e.g. $\mathbf{Set}/T$ defines a setting for multi-set rewriting over the set T. We provide a powerful generalization of this where, through the use of a hierarchy, we can not only guarantee that rewriting an object always returns a well-typed result but, additionally, can dynamically modify the typing object T. Our approach is related to the change-of-base functor familiar from algebraic topology and to its right adjoint whose existence characterizes pullback complements [5]. Indeed, in a sense, our work can be seen as providing a means of exploiting this theory, in a form that can be used for knowledge representation and graph databases, even when only those PBCs required for SqPO rewriting exist.

The arrows in our hierarchies correspond intuitively to the type, or instance-of, relationships found in entity-relationship (ER) modelling [3] or UML, i.e. they are relations that cross from one meta-model layer to another. They also generally correspond to TBox statements in Description Logic [1] although, in some cases, this intuition breaks down since an object, such as the nugget graph of KAMI, with no incoming arrows usually corresponds to a collection of ABox statements about instances of the concepts defined below it in the hierarchy. In this paper, we do not consider the specialization/generalization, or is-a, relation-ships found in ER modelling for the reason that the rewrite of an object does not need to propagate across such relations.

2 Preliminaries

Let us begin by defining a piece of useful terminology. We use the term *element* to refer to any concrete constituent of an object in a concrete category of interest to us, e.g. an element (in the usual sense) of a set or a node or edge of a graph.

2.1 Sesqui-Pushout Rewriting

Sesqui-pushout (SqPO) rewriting [4] is a generalization of double pushout (DPO) and single pushout (SPO) rewriting that allows for the expression of rules for all elementary manipulations generally considered in traditional graph (or multi-set) rewriting: the addition, deletion, merging and cloning of elements. SqPO rewriting can be performed in any category with all pullbacks (PBs), all (final) pullback complements (PBCs) [5] and all pushouts (POs); we further require that POs preserve monos. These conditions are satisfied in all concrete settings of interest to us, typically sets and (simple) graphs with attributes.

In order to perform SqPO rewriting, we only actually need POs of spans where one arrow is a mono. However, in this paper, we sometimes have need of more general POs. We also need the existence of all *image factorizations* (IFs). As this notion is not standard in graph rewriting, we give an explicit definition of its universal property (UP): the image factorization of an arrow $f : A \to B$ is a mono $m : I \rightarrowtail B$ such that (i) there exists an arrow $e : A \to I$ such that $f = m \circ e$; and (ii) for any arrow $e' : A \to I'$ and mono $m' : I' \rightarrowtail B$ such that $f = m' \circ e'$, there exists a unique arrow $i : I \to I'$ such that $m = m' \circ i$

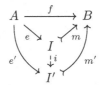

In the concrete settings of interest to us, the IF of an arrow coincides with the familiar notion of its epi–mono factorization. However, we have no (abstract)

need for the first arrow to be an epi and so prefer the more abstract requirement of having IFs of all arrows.

We consider a *rule* to be simply an arrow. A *restrictive instance* of a rule $r^- : L \leftarrow P$ in G is a mono $m : L \rightarrowtail G$ from the target object L; in this case, we refer to L as the LHS and P as the RHS. An *expansive instance* of a rule $r^+ : P \rightarrow R$ is a mono from the source object P; in this case, we refer to P as the LHS and R as the RHS.

The usual notion of rule, i.e. a span of arrows, consists of two rules together with a restrictive instance of the first; the PBC of r^- and m provides an expansive instance m^- of the second and the PO of m^- and r^+ completes the overall rewrite.

$$
\begin{array}{ccccc}
L & \xleftarrow{\;r^-\;} & P & \xrightarrow{\;r^+\;} & R \\
\downarrow{\scriptstyle m} & & \downarrow{\scriptstyle m^-} & & \downarrow{\scriptstyle m^+} \\
G & \xleftarrow[g^-]{} & G^- & \xrightarrow[g^+]{} & G^+
\end{array}
$$

2.2 Hierarchies of Graphs

A *hierarchy* is a finite category $\mathcal{H}$ freely generated from a DAG. An *instance* of a hierarchy is a functor from $\mathcal{H}$ to a category $\mathbf{C}$. This is equivalent to the concrete definition as a DAG. For example, the category generated by $G \rightarrow T$ can be instantiated into **Set** by assigning sets to the two objects and a function between them; such a hierarchy is an *intensional* representation of a multi-set, where G represents all the individuals and T represents the types of individuals, as opposed to an *extensional* representation which would be a function from T to $\mathbf{N}$ assigning multiplicities to types.

In the next two sections, we explain how (i) an expansive rewrite of G is *propagated* to T in order to obtain a rewritten hierarchy $h^+ : G^+ \rightarrow T^+$; and (ii) a restrictive rewrite of T is propagated to G so as to obtain $h^- : G^- \rightarrow T^-$. In the case where $\mathbf{C}$ is **Set**, an expansive rewrite of G applies to the multi-set but propagates to the support of the multi-set if elements of a completely new type are added by the rewrite. On the other hand, a restrictive rewrite of T applies to the support of the multi-set but propagates to the multi-set if a type is deleted from the support. This generalizes standard multi-set rewriting which only ever applies to G and never propagates to T.

In Sect. 5, we extend this to arbitrary hierarchies provided that appropriate *composability* conditions are satisfied and, in Sect. 6, we discuss briefly the current implementation of our framework in the `ReGraph` Python library which works for the more general setting of simple graphs with attributes.

3 Forward Propagation

Throughout this section and the next, we fix two objects G and T and an arrow $h : G \rightarrow T$. In this section, we consider a rule $r : L \rightarrow L^+$ and an expansive instance $m : L \rightarrowtail G$ of r in G. Note that we immediately obtain a typing of L by T by composition, i.e. $h \circ m : L \rightarrow T$.

3.1 The Strict Phase of Forward Rewriting

In order to decide how to propagate a rewrite of G to T, we must further specify to what extent we wish to consider the RHS L^+ of r to be typed by T. There are two extreme cases: the first is where we provide an arrow from L^+ to T, i.e. L^+ is itself typed by T; the other is the case where nothing in the complement of the image of r is homomorphic to T. In the first case, which we call a *strict* rewrite of G, the rewritten G^+ is still typed by T; in the other case, which we call the *canonical* propagation to T, we must propagate all changes in G to T. In between these extremes, we must specify those elements, not in the image of r, that we nonetheless wish to be typed by T.

Definition. Given a rule $r : L \rightarrow L^+$, a *forward factorization* of r is an object L' and arrows $r' : L \rightarrow L'$ and $r^+ : L' \rightarrow L^+$ such that $r = r^+ \circ r'$; and an arrow $x : L' \rightarrow T$ such that $h \circ m = x \circ r'$.

$$
\begin{array}{ccc}
L & \xrightarrow{\ r\ } & L^+ \\
{\scriptstyle hom}\downarrow & {\scriptstyle r'}\searrow & \uparrow{\scriptstyle r^+} \\
T & \xleftarrow{\ x\ } & L'
\end{array}
\tag{1}
$$

In the case of strict rewriting, L' is isomorphic to L^+ so that $x : L^+ \rightarrow T$ whereas, if L' is isomorphic to L, x specifies nothing more than $h \circ m$. In the concrete settings of multi-sets of and graphs, r' is frequently taken to be a mono, i.e. it expresses a rule that only *adds* elements that can be typed by T, but in the abstract setting we have no need to enforce this as a requirement.

The factorization of r splits its application into two phases: the *strict* phase, which modifies only G, and the *canonical* phase, which modifies G and T.

Definition. The *strict rewrite* of G is defined by taking the PO of m and r'. By the definition (1) of forward factorization and the universal property of this PO, we obtain a (unique) arrow h' that types G' by T. (Note that $x = h' \circ m'$.)

$$
\begin{array}{ccc}
L & \xrightarrow{\ r'\ } & L' \\
{\scriptstyle m}\downarrow & & \downarrow{\scriptstyle m'} \\
G & \xrightarrow{\ g'\ } & G' \\
& {\scriptstyle h}\searrow & \downarrow{\scriptstyle h'} \\
& & T
\end{array}
\quad \Big)\, x
\tag{2}
$$

This strict phase of rewriting was discussed briefly in [7] as being the only kind of rewrite that can be performed if T is hard-wired as the base object of a slice category; typically, a descriptive update that *preserves* the current schema.

3.2   The Canonical Phase of Forward Propagation

Our more general and flexible setting of hierarchies enables a second phase of rewriting where the remaining changes to be made to G', as specified by r^+, are additionally propagated to T, i.e. the base object changes.

Definition. The *rewrite* of G is completed by taking the PO of r^+ and m'. The *forward propagation* to T is then defined by taking the PO of g^+ and h'. The final typing of G^+ by T^+ is given by h^+.

$$
\begin{array}{ccc}
L' \xrightarrow{\ r^+\ } L^+ \\
m' \downarrow \qquad \downarrow m^+ \\
G' \xrightarrow[\ g^+\ ]{} G^+
\end{array}
\qquad
\begin{array}{ccc}
G' \xrightarrow{\ g^+\ } G^+ \\
h' \downarrow \qquad \downarrow h^+ \\
T \xrightarrow[\ t^+\ ]{} T^+
\end{array}
$$

$$(3a,3b)$$

Note that we could instead have constructed T^+ by taking the PO of r^+ and $x = h' \circ m'$ and applying the UP of G^+ to construct h^+; the two approaches are equivalent by the pasting lemma for POs. Note also that any object previously typed by G is still typed by G^+ so there is no need to propagate backwards.

The propagated rewrite $t^+ : T \to T^+$ performs all the additions and merges, as specified by r^+ for G', in T to produce the new type T^+ required for G^+. We can also obtain this rewrite by constructing a new rule applying directly to T.

Definition. The *projection* $\hat{r}^+ : L_T \to L_T^+$ of r^+ to T is computed by taking the IF of $h' \circ m'$ followed by the PO of r^+ and $\hat{h}'$. It immediately has an expansive instance $\hat{m}'$ in T.

$$
\begin{array}{ccc}
L' \xrightarrow{\ r^+\ } L^+ \\
\end{array}
$$

$$(4)$$

It is easy to show, by the pasting lemma for POs, that these two definitions of T^+ coincide; as such, for the simple hierarchy $h : G \to T$, we can use either. However, in a general hierarchy where T may be typed by further objects, we must compute the rule projection explicitly in order to continue propagation.

3.3   The Forward Clean-Up Phase

The strict phase of rewriting allows us to add elements to G that can already be typed by T. We may nonetheless fail to include everything that we could have in this strict phase—by inadvertence or simply because, at the time of writing the rule r, we were not aware that some added element could be typed by T.

In order to accommodate this situation, we allow the specification of a *clean-up* phase of rewriting, that applies only to T^+ (and not G^+), by providing an epi $r^\oplus : L_T^+ \twoheadrightarrow L_T^\oplus$; this allows us in particular to merge a newly-added element with another that already existed in T. However, this requires us to know L_T^+—which is dependent on the typing $h' : G' \to T$ and so cannot be specified statically, at the same time as r, but rather dynamically when r's rewrite is propagated to T. The overall effect of such a clean-up is *as if* the original rewrite had specified more in r' and continued with a reduced r^+ and clean-up phase.

Definition. The clean-up phase is specified by an epi $r^\oplus : L_T^+ \twoheadrightarrow T^\oplus$ and the expansive instance $\hat{m}^+ : L_T^+ \rightarrowtail T^+$, obtained after the rewrite of T with the rule projection $\hat{r}^+$ above, giving rise to the final retyping $t^\oplus \circ h^+ : G^+ \to T^\oplus$ of G^+.

$$
\begin{array}{ccccc}
G^+ & & L_T^+ & \xrightarrow{\;r^\oplus\;} & L_T^\oplus \\
 & \searrow^{h^+} & \downarrow{\hat{m}^+} & & \downarrow{\hat{m}^\oplus} \\
 & & T^+ & \xrightarrow[\;t^\oplus\;]{} & T^\oplus
\end{array}
\tag{5}
$$

We ask for $r^\oplus$ to be an epi because, in all concrete models of interest to us, this corresponds to a rule that *only* merges nodes of T^+ in the image of $\hat{m}^+$ and we have no use case for using clean-up to add new elements to T^+.

3.4   Example

Let us illustrate the above theory in a case where G and T are sets, i.e. the hierarchy represents a multi-set. The object T has three elements—white circle, black circle and square—and G has two instances of each of the circles (we use this colour coding to avoid specifying the homomorphisms explicitly). The rule specifies (i) the *merge* of one white and one black circle; and (ii) the addition of a square. The fact that the square can be typed in T is expressed by the arrow from L' to T.

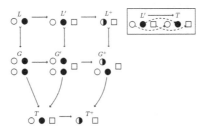

The strict phase of rewriting adds a square to G; the second phase performs the merge in G and propagates this to T. Note that this operation has the *side-effect* that the two circles of G not directly concerned by the rewrite have nonetheless been retyped in T^+.

An alternative factorization of the same rule could have a trivial strict phase and propagate both the merge and the addition to T; in order to arrive at the

same result (up to isomorphism) as the first factorization, we must apply a
clean-up rule to T^+ to merge the black and white squares.

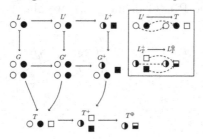

4   Backward Propagation

In this section, we consider the same hierarchy with a rule $r : L \leftarrow L^-$ and a
restrictive instance $m : L \rightarrowtail T$ in T. We can immediately compute the PB of h
and m to obtain a span $\hat{m} : G \leftarrowtail L_G \rightarrow L : \hat{h}$ from the object L_G that can be
seen as the sub-object of G whose *typing* can be modified by r.

$$
\begin{array}{ccc}
L_G & \xrightarrow{\hat{h}} & L \\
{\scriptstyle\hat{m}}\big\downarrow & & \big\downarrow{\scriptstyle m} \\
G & \xrightarrow{h} & T
\end{array}
\tag{6}
$$

4.1   The Strict Phase of Backward Rewriting

As for forward propagation, we must provide a factorization of r in order to
specify which changes to T are to be propagated to G.

Definition. Given a rule $r : L \leftarrow L^-$, a *backward factorization* of r is an object
L' and arrows $r' : L \leftarrow L'$ and $r^- : L' \leftarrow L^-$ such that $r = r^- \circ r'$; and an
arrow $\hat{h}' : L_G \rightarrow L'$ such that $\hat{h} = r' \circ \hat{h}'$. Note that L_G (not G) plays the role
analogous to T in forward propagation.

$$
\begin{array}{ccc}
L & \xleftarrow{\ r\ } & L^- \\
{\scriptstyle\hat{h}}\big\uparrow & \diagup{\scriptstyle r'} & \big\downarrow{\scriptstyle r^-} \\
L_G & \xrightarrow{\ \hat{h}'\ } & L'
\end{array}
\tag{7}
$$

The factorization of r splits its application into two phases: the *strict* phase,
which modifies only T, and the *canonical* phase, which modifies G and T. As
such, in the strict phase of restrictive rewriting, G and L_G remain invariant.
This typically occurs during *concept refinement* where an element of T is cloned
and *all* its instances in G are reassigned a *unique* type in T'.

Definition. The *strict rewrite* of T is defined by taking the PBC of r' and m. By the definition (7) of backward factorization and an application of the UP of the PBC (8a) to the PB (6) defining L_G, we obtain the retyping of G as $h' : G \to T'$.

$$
\begin{array}{ccc}
L & \xleftarrow{\ r'\ } & L' \\
{\scriptstyle m}\downarrow & & \downarrow{\scriptstyle m'} \\
T & \xleftarrow{\ t'\ } & T'
\end{array}
\qquad
\begin{array}{c}
L_G \\
{\scriptstyle \hat{m}}\downarrow \ \ \searrow^{\hat{h}'} \\
G \cdots \ \hat{h} \ \cdots \\
\end{array}
$$

(8a,8b)

Note that any element of T that is deleted must have *no instances* in G for this to be possible—this is a consequence of the requirement that $\hat{h} = r' \circ \hat{h}'$.

4.2   The Canonical Phase of Backward Propagation

Definition. The *rewrite* of T is completed by taking the PBC of r^- and m'. The *backward propagation* to G is then defined by taking the PB of h' and t^-. The final typing of G^- by T^- is simply h^-.

$$
\begin{array}{ccc}
L' & \xleftarrow{\ r^-\ } & L^- \\
{\scriptstyle m'}\uparrow & & \uparrow{\scriptstyle m^-} \\
T' & \xleftarrow{\ t^-\ } & T^-
\end{array}
\qquad\qquad
\begin{array}{ccc}
G & \xleftarrow{\ g^-\ } & G^- \\
{\scriptstyle h'}\downarrow & & \downarrow{\scriptstyle h^-} \\
T' & \xleftarrow{\ t^-\ } & T^-
\end{array}
$$

(9a,9b)

This construction is analogous to the direct construction of T^+ as a PO in forward propagation. If the strict phase of rewriting is trivial, i.e. $L \cong L'$, this corresponds exactly to the notion of (backward) propagation defined in [7]. Note that any object typing T still types T^- so there is no need to propagate forwards.

The propagated rewrite $g^- : G \leftarrow G^-$ performs all the clones and deletions, as specified by r^- for T', in G to produce the new object G^- typed by T^-. We can also obtain this rewrite by constructing a new rule applying directly to G.

Definition. The *lifting* $\hat{r}^- : L_G \leftarrow L_G^-$ of r^- to G is computed by taking the PB of $\hat{h}'$ and r^-. It immediately has the restrictive instance $\hat{m}$, from (6), in G.

$$
\begin{array}{ccc}
L_G & \xleftarrow{\ \hat{r}^-\ } & L_G^- \\
{\scriptstyle \hat{h}'}\downarrow & & \downarrow{\scriptstyle \hat{h}^-} \\
L' & \xleftarrow{\ r^-\ } & L^-
\end{array}
$$

(10)

In this case, we must construct the new typing of G^- by T^- by applying the pasting lemma for PBs (11a) and the UP of the PBC defining T^- (11b) to obtain $h^- : G^- \to T^-$.

$$
\begin{array}{ccc}
L' \xleftarrow{\hat{h}'} L_G \xleftarrow{\hat{r}^-} L_G^- \\
m' \downarrow \qquad \downarrow \hat{m} \qquad \downarrow \hat{m}^- \\
T' \xleftarrow{h'} G \xleftarrow{g^-} G^-
\end{array}
\qquad
\begin{array}{c}
L_G^- \\
m^- \downarrow \quad \searrow^{\hat{h}' \circ \hat{r}^-} \quad \searrow^{\hat{h}^-} \\
G^- \qquad L' \xleftarrow{} L^- \\
{}_{h' \circ g^-}\searrow \quad h^- \downarrow \qquad \downarrow \\
T' \xleftarrow{} T^-
\end{array}
\qquad \text{(11a,11b)}
$$

The proof of the equivalence of the two definitions of G^- is a little more complex than for its analogue for forward propagation; it requires an application of the UP of the PBC (9a) defining T^- followed by two applications of the UP of the PB (9b) defining G^-. As for forward propagation, we must compute the rule lifting explicitly ii G itself types further graphs in the original hierarchy.

Note how the strict phase of rewriting allows us to 'protect' elements of G from being cloned; instead, it retypes those elements with the more refined type T'—a process called *concept refinement* or *specialization* in ER modelling [3]—and clones only the remaining nodes, as specified by r^-.

4.3   The Backward Clean-Up Phase

We can specify a clean-up phase of rewriting, only for G^-, by providing a mono $r^\ominus : L_G^- \leftarrow L_G^\ominus$. Clearly, and analogously to the situation for forward propagation, in order to provide such an $r^\ominus$, we already need to know L_G^-—which is dependent on the typing $G \to T'$. As such, $r^\ominus$ cannot be specified statically, at the same time as the original r used to rewrite T, but should rather be provided dynamically at the time that r's rewrite is being propagated to G.

$$
\begin{array}{ccc}
L_G^- & \xleftarrow{\;r^\ominus\;} & L_G^\ominus \\
\hat{m}^- \downarrow & & \downarrow \hat{m}^\ominus \\
G^- & \xleftarrow{\;g^\ominus\;} & G^\ominus \\
h^- \downarrow & & \\
T^- & &
\end{array}
\qquad \text{(12)}
$$

The clean-up phase allows us to remove undesired element clones that were not specified during the strict phase of rewriting, e.g. a *partial* concept refinement where some instances of a cloned element cannot be assigned a unique type in T'. However, if $r^\ominus$ is not a mono, this phase can also create additional clones, beyond what was specified by r, and we have no use case for this extra generality, just as we have no use case for allowing the clean-up phase to add new elements to T^+ during forward propagation.

4.4   Example

We again consider an example where our hierarchy represents a multi-set. The rule specifies (i) the deletion of the circle; and (ii) the cloning of the square into a (white) square and a black square. The fact that one square in G is to become white while the other becomes black is expressed by the arrow from L_G to L'.

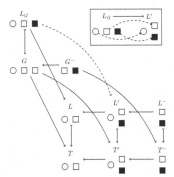

The strict phase of rewriting clones the square and retypes G, thus effecting a concept refinement; the second phase deletes the circle and propagates to G.

If we have a third instance of the square in G for which we *cannot* assign a unique new type in T', we must displace the cloning operation to the second phase of rewriting and propagate to *all* instances of the square. In order to recover the same retyping of (the first two) squares as above, we must apply a clean-up rule to delete the unwanted clones.

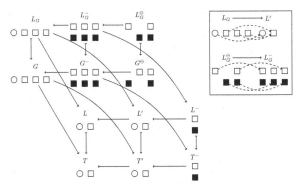

5   Rewriting General Hierarchies

The notions of forward and backward propagation described in the previous two sections enable the unambiguous rewrite of a tree (or forest) hierarchy. However, in a hierarchy containing an undirected cycle

$$
\begin{array}{c}
G \\
{}^{h_1}\swarrow \quad \Big\downarrow {}^{h_2} \\
T_1 \\
{}_{h_{12}}\searrow \quad \downarrow \\
T_2
\end{array}
\tag{13}
$$

an expansive rewrite of G could, in general, propagate incoherently to T_1 and T_2 in the sense that we cannot reconstruct a (unique) arrow $h_{12}^+ : T_1^+ \to T_2^+$ making the triangle commute.

Similarly, a restrictive rewrite of T in a hierarchy such as

$$
\begin{array}{c}
G_1 \\
{}^{h_{12}}\swarrow \quad \Big\downarrow {}^{h_1} \\
G_2 \\
{}_{h_2}\searrow \quad \downarrow \\
T
\end{array}
\tag{14}
$$

may propagate incoherently to G_1 and G_2.

In this section, we characterize the conditions under which the arrows h_{12} can be reconstructed in the rewritten hierarchy. We provide only the overall structure of the argument, for reasons of space, but full details will be given in the full version of this paper. We only need to consider the cases of the two triangles above as all hierarchies can be decomposed into multiple such triangles.

5.1   Forward Composability

Definition. Given a rule $r : L \to L^+$, an expansive instance $m : L \rightarrowtail G$ and factorizations

$$
\begin{array}{ccc}
L \xrightarrow{\;r\;} L^+ & \qquad & L \xrightarrow{\;r\;} L^+ \\
{}_{h_1 \circ m}\Big\downarrow \;{}_{r_1'}\searrow\; \Big\uparrow{}^{r_1^+} & \qquad & {}_{h_2 \circ m}\Big\downarrow \;{}_{r_2'}\searrow\; \Big\uparrow{}^{r_2^+} \\
T_1 \xleftarrow{\;x_1\;} L_1' & \qquad & T_2 \xleftarrow{\;x_2\;} L_2'
\end{array}
\tag{15a,15b}
$$

that define the propagation of r to T_1 and T_2 respectively, we say that the two propagations $h_1^+ : G^+ \to T_1^+$ and $h_2^+ : G^+ \to T_2^+$ are *composable* iff there exists a unique $h_{12}^+ : T_1^+ \to T_2^+$ such that

$$
\begin{array}{ccc}
G & \longrightarrow & G^+ \\
\swarrow \;\Big\downarrow & & \swarrow \;\Big\downarrow \\
T_1 \longrightarrow T_1^+ & & {}_{h_{12}^+}\Big\downarrow \\
\searrow \;\downarrow & & \searrow \;\downarrow \\
T_2 & \longrightarrow & T_2^+
\end{array}
\tag{16}
$$

In order to construct h_{12}^+, we need to apply the UP of the PO (3b) defining T_1^+ according to (17a) below. We can zoom in on the precise conditions that must be satisfied to apply this UP by superimposing the span of the PO defining T_1^+ with the PO defining T_2^+, as in (17b).

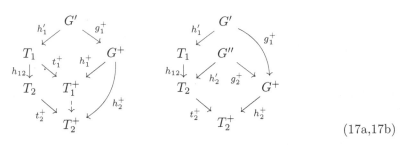

$$(17a,17b)$$

If we can construct an arrow $g : G' \to G''$ such that

$$(18)$$

then we can indeed apply the UP, as in (17a), and obtain a unique $h_{12}^+ : T_1^+ \to T_2^+$ satisfying the composability conditions.

The existence of a *unique* such g follows from the existence of an arrow $\ell : L_1' \to L_2'$ satisfying

$$(19a,19b)$$

The proof is elementary and follows from two applications—to G'' and G^+—of the UP of the PO defining G'. The existence of such an ℓ means that all add and merge operations performed by r_1' are also performed by r_2'; otherwise g need not be unique and may not even exist.

If we further perform clean-up phases, we must additionally ask for a unique $\ell^\oplus : L_1^\oplus \to L_2^\oplus$ such that

$$
\begin{array}{ccc}
L_1^+ & \xrightarrow{r_1^\oplus} & L_1^\oplus \\
\downarrow{\ell^+} & & \vdots{\ell^\oplus} \\
L_2^+ & \xrightarrow{r_2^\oplus} & L_2^\oplus
\end{array}
\qquad (20)
$$

where $\ell^+ : L_1^+ \to L_2^+$ is defined by the UP of the PO (4) defining the rule projection of r_1 and therefore satisfies $h_{12}^+ \circ \hat{m}_1^+ = \hat{m}_2^+ \circ \ell^+$. We can then apply the UP of the PO (5) defining $T_1^\oplus$ to obtain a unique arrow $h_{12}^\oplus : T_1^\oplus \to T_2^\oplus$.

Proposition. Given the forward factorizations (15a,15b) and $\ell : L_1' \to L_2'$ satisfying (19a,19b), there is a unique $h_{12}^+ : T_1^+ \to T_2^+$ satisfying (16); and given an $\ell^\oplus : L_1^\oplus \to L_2^\oplus$ satisfying (20), there is a unique $h_{12}^\oplus$ such that $t_2^\oplus \circ h_{12}^\oplus = h_{12}^\oplus \circ t_1^\oplus$.

5.2 Backward Composability

Definition. Given a rule $r : L \leftarrow L^-$, a restrictive instance $m : L \rightarrowtail T$ and factorizations

$$
\begin{array}{ccc}
L \xleftarrow{\;r\;} L^- & \qquad & L \xleftarrow{\;r\;} L^- \\
\hat{h}_1 \uparrow \quad \nwarrow^{r_1'} \quad \downarrow r_1^- & & \hat{h}_2 \uparrow \quad \nwarrow^{r_2'} \quad \downarrow r_2^- \\
L_1 \xrightarrow[\hat{h}_1']{} L_1' & & L_2 \xrightarrow[\hat{h}_2']{} L_2'
\end{array}
\tag{21a,21b}
$$

that define the propagation of r to G_1 and G_2 respectively (where L_i is shorthand for L_{G_i} as defined by the PB (6)), the two propagations $h_1^- : G_1^- \to T^-$ and $h_2^- : G_2^- \to T^-$ are *composable* iff there exists a unique $h_{12}^- : G_1^- \to G_2^-$ such that

$$
\begin{array}{ccc}
G_1 \longleftarrow & & G_1^- \\
\swarrow \quad\;\; {\scriptstyle h_{12}^-} \nearrow & & \big\downarrow \\
G_2 \longleftarrow G_2^- & & \\
\searrow \quad \big\downarrow \quad \searrow & & \big\downarrow \\
T \longleftarrow & & T^-
\end{array}
\tag{22}
$$

Analogously but dually to the case of forward composability, we can construct the desired h_{12}^- by applying the UP of the PB (9b) defining G_2^-, as in (23a).

$$
\tag{23a,23b}
$$

In order to prove that the outer square commutes, we therefore need an arrow $t : T' \to T''$ satisfying

$$
\begin{array}{ccc}
 & T' \xleftarrow{\;h_1'\;} G_1 \\
{\scriptstyle t_1^-} \nearrow \;\;\big\uparrow & & \big\downarrow {\scriptstyle h_{12}} \\
T^- \quad \big\downarrow{\scriptstyle t} & & \\
{\scriptstyle t_2^-} \searrow \;\; \big\downarrow & & \big\downarrow \\
 & T'' \xleftarrow{\;h_2'\;} G_2
\end{array}
\tag{24}
$$

We first apply the UP of the PB (6) defining L_2 to obtain a unique arrow $\hat{h}_{12} : L_1 \to L_2$ satisfying

$$\tag{25}$$

By the inverse pasting lemma for PBs, the resulting square is moreover a PB. The existence of a *unique* such $t : T' \to T''$ now follows from the existence of an arrow $\ell : L_1' \to L_2'$ satisfying

$$\tag{26a,26b}$$

The proof is elementary and follows from three applications of the UP of the PBC defining T''—to T', L_1 and T^-.

For the clean-up phase, we apply the UP of the PB defining L_2^- to obtain an arrow $\ell^- : L_1^- \to L_2^-$ and further require a unique arrow $\ell^\ominus : L_1^\ominus \to L_2^\ominus$ such that $r_2^\ominus \circ \ell^\ominus = \ell^- \circ r_1^\ominus$. We then apply the UP of the PBC defining $G_2^\ominus$ to obtain the desired arrow $h_{12}^\ominus : G_1^\ominus \to G_2^\ominus$.

Proposition. Given the backward factorizations (21a,21b) and an $\ell : L_1' \to L_2'$ satisfying (26a,26b), there is a unique $h_{12}^- : G_1^- \to G_2^-$ satisfying (22); and given an $\ell^\ominus : L_1^\ominus \to L_2^\ominus$ satisfying $r_2^\ominus \circ \ell^\ominus = \ell^- \circ r_1^\ominus$, there is a unique $h_{12}^\ominus : G_1^\ominus \to G_2^\ominus$ satisfying $h_{12}^- \circ g_1^\ominus = g_2^\ominus \circ h_{12}^\ominus$.

6   The ReGraph Library

The ReGraph Python library[1] implements general hierarchies of simple graphs with attributes. It supports both in-memory graphs, via the networkX library, and persistent graphs, via the Neo4j graph database. Rules can be expressed declaratively, essentially using the mathematical definition used in this paper, or procedurally, using a simple language to express the primitive operations of clone, delete, add and merge.

The principal difference between the theory presented in this paper and the implementation lies in the specification of propagation: in ReGraph, a controlled propagation is specified by a single relation that plays the same rôle as the strict and clean-up phases presented here. For example, the partial concept refinement of Sect. 4.4 is expressed by the same rule together with a relation that specifies,

---

[1] https://github.com/Kappa-Dev/ReGraph.

for the first two squares, how to retype them; nothing need be specified for the third square which, as a result, is cloned. This alternative means of specifying propagation can be formulated mathematically and shown to be equivalent to the phased propagation presented in this paper; we will provide the details in the full version of this paper.

The bio-curation tool KAMI[2] [8], discussed in the introduction, is based on the ReGraph library. It makes extensive use of forward propagation, in order to aggregate new PPIs appropriately into an existing knowledge corpus, e.g. if it identifies that a node mentioned in an input already exists in the action graph, it constructs a strict rewrite, to reuse that node, rather than creating a new one by canonical propagation. It also makes use of backward propagation in order to contextualize knowledge to a particular collection of gene products. Indeed, these were the original, informal use cases of propagation which motivated the development of the theory presented in this paper.

The implementation[3] of evolvable schemas for Neo4j [2] also uses ReGraph; this implements the translation of the procedural language for specifying rules into OpenCypher. In this setting, forward propagation constructs an automatic update of the schema graph in the light of an update of the data graph that would otherwise break schema validation, i.e. a descriptive update. Dually, backward propagation constructs an automatic update of the data graph in the event of an update of the schema, i.e. a prescriptive update. In order to build a fully general front-end to Neo4j in this way, we need to extend ReGraph to work with non-simple graphs. This poses no conceptual problem and the theory of propagation described here applies without any changes to this more expressive setting.

7   Conclusions

We have presented a formalism for graph-based knowledge representation and update that exploits SqPO rewriting to perform updates anywhere in a hierarchy of objects (typically sets or graphs). For this extended abstract, we have chosen a rigorous, but largely informal, presentation; the main contribution of the paper can be stated as follows:

Given a hierarchy, a SqPO rule, an expansive (resp. restrictive) instance of that rule into some object O and factorizations for *every* other object on a path from (resp. to) O satisfying composability, we can *uniquely* rewrite the entire hierarchy in a way that guarantees the validity of the result.

The requirement to specify all these factorizations—and also verify that they satisfy composability if necessary—can, in principle, be very onerous. Nonetheless, our experience suggests that most updates need propagate only along single edges or, at most, paths of length 2 so that, in practice, the requirement is not too onerous.

The other principal open question concerns the characterization of the data structures necessary to maintain an *audit trail* of all updates made to a system.

---

[2] https://github.com/Kappa-Dev/KAMI.

[3] https://github.com/Kappa-Dev/ReGraph/blob/master/regraph/neo4j/graphs.py.

This would enable us to determine whether an update can be undone or not, a question that is greatly complicated by the fact of propagation, and, more generally, provide support for maintaining different *versions* of the contents of a KR. This requires a major generalization of the theory of *causality* between SqPO rules; see [7] for example. We intend to investigate this question first in the two concrete use cases discussed in this paper before attempting a full-blown generalization to arbitrary hierarchies.

References

1. Baader, F., Calvanese, D., McGuinness, D., Patel-Schneider, P., Nardi, D.: The Description Logic Handbook: Theory, Implementation and Applications. CUP, Cambridge (2003)
2. Bonifati, A., Furniss, P., Green, A., Harmer, R., Oshurko, E., Voigt, H.: Schema validation and evolution for graph databases. arXiv preprint arXiv:1902.06427 (2019)
3. Chen, P.P.S.: The entity-relationship model–toward a unified view of data. ACM Trans. Database Syst. (TODS) 1(1), 9–36 (1976)
4. Corradini, A., Heindel, T., Hermann, F., König, B.: Sesqui-pushout rewriting. In: Corradini, A., Ehrig, H., Montanari, U., Ribeiro, L., Rozenberg, G. (eds.) ICGT 2006. LNCS, vol. 4178, pp. 30–45. Springer, Heidelberg (2006). https://doi.org/10.1007/11841883_4
5. Dyckhoff, R., Tholen, W.: Exponentiable morphisms, partial products and pullback complements. J. Pure Appl. Algebra 49(1–2), 103–116 (1987)
6. Francis, N., et al.: Cypher: an evolving query language for property graphs. In: Proceedings of the 2018 International Conference on Management of Data, pp. 1433–1445. ACM (2018)
7. Harmer, R.: Rule-based meta-modelling for bio-curation. Habilitation à Diriger des Recherches, ENS Lyon, France (2017)
8. Harmer, R., Cornec, Y.-S.L., Légaré, S., Oshurko, I.: Bio-curation for cellular signalling: the KAMI project. In: Feret, J., Koeppl, H. (eds.) CMSB 2017. LNCS, vol. 10545, pp. 3–19. Springer, Cham (2017). https://doi.org/10.1007/978-3-319-67471-1_1

Relating DNA Computing
and Splitting/Fusion Grammars

Hans-Jörg Kreowski, Sabine Kuske, and Aaron Lye$^{(\boxtimes)}$

Department of Computer Science and Mathematics, University of Bremen,
P.O.Box 33 04 40, 28334 Bremen, Germany
{kreo,kuske,lye}@informatik.uni-bremen.de

Abstract. Splitting/fusion grammars were recently introduced as devices for the generation of hypergraph languages. Their rule application mechanism is inspired by basic operations of DNA computing. In this paper, we demonstrate that splitting/fusion grammars and well-known computational approaches based on DNA computing are closely related on a technical level beyond the mere motivation. This includes Adleman's seminal experiment, insertion-deletion systems, and extended iterated 2-splicing systems.

1 Introduction

Adleman demonstrated in his seminal experiment [1] that the NP-hard Hamiltonian path problem can be solved by a polynomial number of biochemical operations on DNA strands with high probability exploiting the parallelism of chemical reactions in tubes of molecules. This was the starting point of the area of DNA computing that has been intensely developed since then. Inspired by DNA computing, we introduced fusion grammars in [2] and splicing/fusion grammars in [3]. In this paper, we rename the latter by splitting/fusion grammars as the term "splicing" may be misleading.

The core of DNA computing is the biochemical processing on tubes of DNA molecules. In the framework of splitting/fusion grammars, we exploit similarities between hypergraphs and tubes of molecules, which are multisets from a mathematical point of view. Each hypergraph is the disjoint union of its connected components which corresponds to a multiset if one counts the isomorphic connected components. Therefore, connected components of hypergraphs can be seen as counterparts of DNA molecules. To emphasize this analogy, we call the connected components molecules. Furthermore, we reflect the Watson-Crick complementarity of DNA nucleotides and single DNA strands by a complementarity of hyperedges and the basic DNA operations ligation, restriction, duplication by polymerase chain reaction, and reading by gel electrophoresis by fusion, splitting, multiplication, and filtering of special connected components, respectively. In this paper, we show that the relation between DNA computing and splitting/fusion grammars goes far beyond mere motivation. We model three well-known DNA computing approaches in our framework. In Sect. 4, we

E. Guerra and F. Orejas (Eds.): ICGT 2019, LNCS 11629, pp. 159–174, 2019.
https://doi.org/10.1007/978-3-030-23611-3_10

recreate Adleman's experiment in terms of fusion grammars. In Sect. 5, insertion-deletion systems as one of the prominent (string) language generating devices based on DNA computing (cf., e.g., Chapter 6 of [4]) are transformed into split-ting/fusion grammars. Another important DNA computing approach offers splicing systems in many variants (cf., e.g., Chapters 7 to 11 of [4]). In Sect. 6, we generalize extended iterated 2-splicing systems to 2-splicing grammars that are special regulated splitting/fusion grammars. Section 2 provides preliminaries for hypergraphs and the notion of splitting/fusion grammars is recalled in Sect. 3. Section 7 concludes the paper. The proofs of all stated correctness results are omitted because of the page limit.

2   Preliminaries

In this section, basic notions and notations of hypergraphs are recalled (see, e.g., [5]).

Let Σ be a label alphabet. A *hypergraph* over Σ is a system $H = (V, E, att, lab)$ where V is a finite set of *nodes*, E is a finite set of *hyperedges*, $att \colon E \to V^*$ is a function, called *attachment* (assigning a string of attachment nodes to each edge), and $lab \colon E \to \Sigma$ is a function, called *labeling*.

The length of the attachment $att(e)$ for $e \in E$ is called *type* of e, and e is called *A-hyperedge* if A is its label. Let $\Sigma' \subseteq \Sigma$ be a subalphabet of Σ and $type \colon \Sigma' \to \mathbb{N}$ a function, called *type function*. Then we require that every A-hyperedge with $A \in \Sigma'$ is of type $type(A)$. The components of $H = (V, E, att, lab)$ may also be denoted by V_H, E_H, att_H, and lab_H respectively. The class of all hypergraphs over Σ is denoted by $\mathcal{H}_\Sigma$.

A *(directed) graph* is a hypergraph $H = (V, E, att, lab)$ with $att(e) \in V^2$ for all $e \in E$. In this case, the hyperedges are called *edges*. If $att(e) = vv$ for some $v \in V$, then e is also called a *loop*. A graph is called *loop-free* if for all $e \in E$ $att(e) = vv'$ with $v \neq v'$. A graph is called *simple* if it is loop-free and no parallel edges exist.

The set $\{1, \ldots, k\}$ for some $k \in \mathbb{N}$ is denoted by $[k]$ which also denotes the discrete graph with the nodes $1, \ldots, k$ and an empty set of hyperedges.

In drawings, an A-hyperedge e with attachment $att(e) = v_1 \cdots v_k$ is depicted

by . Moreover, a hyperedge of type 2 may be depicted as

an edge by instead of . If there are two edges with the same label, but in opposite directions, we may draw them as an undirected edge. We assume the existence of a special label $* \in \Sigma$ that is omitted in drawings. We call a hypergraph *unlabeled* if $lab(e) = *$ for all $e \in E$.

Given $H, H' \in \mathcal{H}_\Sigma$, H is a *subhypergraph* of H' if $V_H \subseteq V_{H'}$, $E_H \subseteq E_{H'}$, $att_H(e) = att_{H'}(e)$, and $lab_H(e) = lab_{H'}(e)$ for all $e \in E_H$. This is denoted by $H \subseteq H'$.

Let $H \in \mathcal{H}_\Sigma$. Then a sequence of triples $(i_1, e_1, o_1) \ldots (i_n, e_n, o_n) \in (\mathbb{N} \times E_H \times \mathbb{N})^*$ is a *path* from $v \in V_H$ to $v' \in V_H$ if $v = att_H(e_1)_{i_1}, v' = att_H(e_n)_{o_n}$

and $att_H(e_j)_{o_j} = att_H(e_{j+1})_{i_{j+1}}$ for $j = 1, \ldots, n-1$ where, for each $e \in E_H$, $att_H(e)_i = v_i$ for $att_H(e) = v_1 \cdots v_k$ and $i = 1, \ldots, k$. In the case of simple graphs, a path may be denoted by the sequence of visited nodes as the involved edges are uniquely determined.

H is *connected* if each two nodes are connected by a path. A connected subgraph M of H is called a *molecule* of H if it is maximal meaning that $M \subseteq M' \subseteq H$ for a connected M' implies $M = M'$. The set of molecules of H is denoted by $\mathcal{M}(H)$.

Given $H, H' \in \mathcal{H}_\Sigma$, a *(hypergraph) morphism* $g \colon H \to H'$ consists of two mappings $g_V \colon V_H \to V_{H'}$ and $g_E \colon E_H \to E_{H'}$ such that $att_{H'}(g_E(e)) = g_V^*(att_H(e))$ and $lab_{H'}(g_E(e)) = lab_H(e)$ for all $e \in E_H$, where $g_V^* \colon V_H^* \to V_{H'}^*$ is the canonical extension of g_V, given by $g_V^*(v_1 \cdots v_n) = g_V(v_1) \cdots g_V(v_n)$ for all $v_1 \cdots v_n \in V_H^*$. H and H' are *isomorphic*, denoted by $H \cong H'$, if there is an isomorphism $g \colon H \to H'$, i.e., a morphism with bijective mappings. Clearly, $H \subseteq H'$ implies that the two inclusions $V_H \subseteq V_{H'}$ and $E_H \subseteq E_{H'}$ define a morphism $incl \colon H \to H'$. Given a morphism $g \colon H \to H'$, the image of H in H' under g defines the subgraph $g(H) \subseteq H'$.

Let $H' \in \mathcal{H}_\Sigma$ as well as $V \subseteq V_{H'}$ and $E \subseteq E_{H'}$. Then the *removal* of (V, E) from H' given by $H = H' - (V, E) = (V_{H'} - V, E_{H'} - E, att_H, lab_H)$ with $att_H(e) = att_{H'}(e)$ and $lab_H(e) = lab_{H'}(e)$ for all $e \in E_{H'} - E$ defines a subgraph $H \subseteq H'$ if $att_{H'}(e) \in (V_{H'} - V)^*$ for all $e \in E_{H'} - E$, i.e., no remaining hyperedge is attached to a removed node. This condition is called *dangling condition*. The dangling condition is fulfilled in the special case that only hyperedges are removed.

Given $H, H' \in \mathcal{H}_\Sigma$, the *disjoint union* of H and H' is denoted by $H + H'$. A special case is the disjoint union of H with itself k times, denoted by $k \cdot H$. Let $H \in \mathcal{H}_\Sigma$ and $m \colon \mathcal{M}(H) \to \mathbb{N}$ be a mapping, called *multiplicity*. Then the multiplication of m and H is defined by $m \cdot H = \sum_{M \in \mathcal{M}(H)} m(M) \cdot M$. The disjoint union is unique up to isomorphism. It is easy to see that the disjoint union is commutative and associative. Moreover, there are injective morphisms $in_H \colon H \to H + H'$ and $in_{H'} \colon H' \to H + H'$ such that $in_H(H) \cup in_{H'}(H') = H + H'$ and $in_H(H) \cap in_{H'}(H') = \emptyset$. Each two morphisms $g_H \colon H \to Y$ and $g_{H'} \colon H' \to Y$ define a unique morphism $\langle g_H, g_{H'} \rangle \colon H + H' \to Y$ with $\langle g_H, g_{H'} \rangle \circ in_H = g_H$ and $\langle g_H, g_{H'} \rangle \circ in_{H'} = g_{H'}$. In particular, one gets $g = \langle g \circ in_H, g \circ in_{H'} \rangle$ for all morphisms $g \colon H + H' \to Y$ and $g + g' = \langle in_Y \circ g, in_{Y'} \circ g' \rangle \colon H + H' \to Y + Y'$ for morphisms $g \colon H \to Y$ and $g' \colon H' \to Y'$. The disjoint union is the coproduct in the category of hypergraphs.

The merging of nodes is defined as a quotient by means of an equivalence relation $\equiv$ on the set of nodes V_H of H as follows: $H/\equiv = (V_H/\equiv, E_H, att_{H/\equiv}, lab_H)$ with $att_{H/\equiv}(e) = [v_1] \cdots [v_k]$ for $e \in E_H$, $att_H(e) = v_1 \cdots v_k$ where $[v]$ denotes the equivalence class of $v \in V_H$ and $V_H/\equiv$ is the set of equivalence classes. Given two sequences $d(1) \cdots d(k)$ and $d'(1) \cdots d'(k)$ of nodes in V_H for some $k \in \mathbb{N}$. Then the relation $d(i) \sim d'(i)$ for $i = 1, \ldots, k$ induces a particular equivalence relation (by the reflexive, symmetric and transitive closure of $\sim$) that is denoted by $d = d'$. This is employed in the next section to define the fusion of hyperedges. It is easy to see that $f: H \to H/\!\!\equiv$ given by $f_V(v) = [v]$ for all $v \in V_H$ and $f_E(e) = e$ for all $e \in E_H$ defines a *quotient morphism*.

3   Splitting/Fusion Grammars

In this section, we recall the notion of splitting/fusion grammars which are called splicing/fusion grammars in [3]. The grammars are renamed because we think that "splitting" fits better. We begin with the concept of fusion considered as application of fusion rules, and then continue with splitting which is converse to fusion.

Definition 1. 1. Let $F \subseteq \Sigma$ be a finite set of labels and $k: F \to \mathbb{N}$ a type function. F is called *fusion alphabet*, its elements *fusion labels*. Let $\overline{a} \in \Sigma$ be a *complementary fusion label* for each $a \in F$ such that $\overline{a} \neq \overline{b}$ for all $a \neq b$. The set of complementary fusion labels is denoted by $\overline{F}$. The typing function and the complementarity are extended to $\overline{F}$ by $k(\overline{a}) = k(a)$ and $\overline{\overline{a}} = a$ for all $\overline{a} \in \overline{F}$. Let $H \in \mathcal{H}_\Sigma$ and $e, \overline{e} \in E_H$ with $a = lab_H(e) = \overline{lab_H(\overline{e})}$ for some $a \in F$. Then the *fusion of e and $\overline{e}$ in H* yields the hypergraph $H_{fuse(e,\overline{e})} = (H - (\emptyset, \{e, \overline{e}\}))/_{att_H(e)=att_H(\overline{e})}$.
2. Let $H, H' \in \mathcal{H}_\Sigma$. Then H *directly derives* H' through fusion wrt $a \in F$ if $H' \cong H_{fuse(e,\overline{e})}$ for some $e, \overline{e} \in E_H$ with $a = lab_H(e) = \overline{lab_H(\overline{e})}$. Here the fusion label $a \in F$ plays the role of a *fusion rule* indicated by the notation $fr(a)$. Its application is denoted by $H \underset{fr(a)}{\Longrightarrow} H'$.

Remark 1. 1. As each hyperedge belongs to a single molecule, the fusion of two hyperedges changes either one molecule or two molecules. Moreover, a fusion can have three different effects.
   – It may be a kind of folding, e.g., $\boxed{a}\!-\!\bullet\!-\!\bullet\!-\!\boxed{\overline{a}} \underset{fr(a)}{\Longrightarrow} \bullet\!\!\multimap$,
   – two molecules may be joint, e.g., 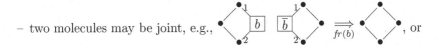 , or
   – it can also result in disconnection, e.g., $\boxed{c}\ \boxed{\overline{c}} \underset{fr(c)}{\Longrightarrow} \bullet$ wrt one molecule

   or $\bullet\!\!\overset{1}{-}\!\boxed{d}\!\overset{2}{-}\!\bullet\ \ \bullet\!\!\overset{1}{-}\!\boxed{\overline{d}}\!\overset{2}{-}\!\bullet \underset{fr(d)}{\Longrightarrow} \bullet\ \ \bullet$ wrt two molecules.
2. It is easy to see that fusion rules can be applied in parallel if their matchings access pairwise different hyperedges (cf. [2]).

Definition 2. 1. Let $H' \in \mathcal{H}_\Sigma$ and $a \in F$. Then $H \in \mathcal{H}_\Sigma$ is a *splitting* of H' wrt a if there are $e, \overline{e} \in E_H$ with $a = lab_H(e) = \overline{lab_H(\overline{e})}$ such that $H' \cong H_{fuse(e,\overline{e})}$.
2. Such a splitting can be considered as a direct derivation $H' \underset{sr(a)}{\Longrightarrow} H$, where $sr(a)$ indicates that the label a is used as a splitting rule.

Remark 2. 1. To get more flexibility, we use other complementary labels for splitting than for fusion. For $a \in F$, the *complementary splitting label of a* is denoted by $\hat{a}$ (instead of $\bar{a}$).

2. An application of $sr(a)$ to H' can be explicitly performed by (1) choosing a *matching morphism* $g': [k(a)] \to H'$, i.e., a sequence of nodes $g'(1) \cdots g'(k(a))$ in H', (2) for $i = 1, \ldots, k(a)$, either splitting $g'(i)$ into two new nodes $d(i)$ and $\hat{d}(i)$ or replacing it by one new node v with $d(i) = v = \hat{d}(i)$, subject to the condition: $g'(i) = g'(j)$ for some $i \neq j$ if and only if $d(i) = d(j)$ and $\hat{d}(i) = \hat{d}(j)$, (3) constructing the hypergraph $I' = (V_{H'} \setminus \{g'(i) \mid i = 1, \ldots, k(a)\} + \{d(i), \hat{d}(i) \mid i = 1, \ldots, k(a)\}, E_{H'}, att_{I'}, lab_{H'})$ with $att_{I'}(e')_j = d(i)$ or $att_{I'}(e')_j = \hat{d}(i)$ for $j = 1, \ldots, k(a)$ provided that $att_{H'}(e')_j = g'(i)$ for some $i = 1, \ldots, k(a)$ and $att_{I'}(e') = att_{H'}(e')$ otherwise, (4) constructing H'' from I' by adding two new hyperedges e and $\hat{e}$ with $att_{H''}(e) = d(1) \cdots d(k(a))$ and $att_{H''}(\hat{e}) = \hat{d}(1) \cdots \hat{d}(k(a))$ as well as $lab_{H''}(e) = a$ and $lab_{H''}(\hat{e}) = \hat{a}$, and (5) renaming nodes and hyperedges of H'' optionally. I' in (3) is called *intermediate hypergraph.*

3. While the application of a fusion rule is unique up to isomorphism, splitting is highly nondeterministic in general because each tentacle of a hyperedge of H' that is attached to a matching node $g'(i)$ may be attached to $d(i)$ or $\hat{d}(i)$ in H''. Consider, for example, a fusion symbol t of type 2 and a triangle

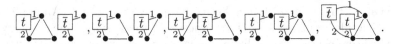 where the numbered nodes define the matching. Although this is a

very small graph with much symmetry, one gets five splittings:

$$\boxed{t}\triangle, \boxed{\hat{t}}\triangle, \boxed{t}\triangle, \boxed{\hat{t}}\triangle, \boxed{t}\triangle\boxed{\hat{t}}, \boxed{t}\triangle\boxed{\hat{t}}, \boxed{t}\triangle.$$

As the high nondeterminism of splitting is not always desirable, we employ some variants of context conditions to cut the nondeterminism down.

Definition 3. Let F be a fusion alphabet and $a \in F$.

1. A *splitting rule with a fixed disjoint subcontext* is a triple $(sr(a), d: [k(a)] \to D, incl: D \to C)$ where $C, D \in \mathcal{H}_\Sigma$ are connected hypergraphs, d is a morphism, and $incl$ is an injective morphism. It can be applied to H' wrt the matching morphism $g': [k(a)] \to H'$ if there is a morphism $f: C \to H'$ and the intermediate hypergraph I' can be chosen as $I'' + D$ with $d(i) \in V_{I''}$ and $\hat{d}(i) \in V_D$ for $i = 1, \ldots, k(a)$ and $I'' \subseteq H'$ such that $g' = d \circ incl \circ f$.

2. A *splitting rule with double context* is a triple $(sr(a), c_1: [k(a)] \to C_1, c_2: [k(a)] \to C_2)$ where C_1 and C_2 are connected hypergraphs and c_1 and c_2 morphisms. It is applicable to H' wrt the matching morphism $g': [k(a)] \to H'$ if there are morphisms $f_j: C_j \to H', j = 1, 2$, with $f_1(C_1) \cap f_2(C_2) = g'([k(a)]) = f_1(c_1([k(a)])) = f_2(c_2([k(a)]))$ and the intermediate hypergraph I' can be chosen as $I_1 + I_2$ such that there are morphisms $f'_j: C_j \to I_j$ for $j = 1, 2$ and injective morphisms $in_j: I_j \to H'$ for $j = 1, 2$ such that $in_j \circ f'_j = f_j$ for $j = 1, 2$.

Fig. 1. Application of a splitting rule $r = (sr(a), d, incl)$ with fixed disjoint subcontext where $E = (f(C) - f(D)) + f([k(a)])$

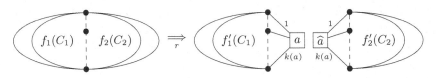

Fig. 2. Application of a splitting rule with double context $r = (sr(a), c_1, c_2)$

Remark 3. 1. The splitting rules with fixed disjoint subcontext are used in Sect. 5 together with a variant for the special types of graphs considered there. Splitting rules with fixed disjoint subcontext where context and subcontext coincide were already used in [3]. While the context is required to be present in the processed hypergraphs, the subcontext is required to be a disjoint component after splitting. Figure 1 illustrates such a splitting for a single molecule. This is not always possible as there may be a hyperedge in the processed hypergraphs that does not belong to the matching of the subcontext, but one of its tentacles is attached to an inner node of the subcontext, i.e., a node that is not matched by g'.

2. The splitting rules with double context are used in Sect. 6. An application of such a rule is only possible if the processed hypergraph can be cut in two parts which intersect in the splitting nodes only and where one context matches in one part and the other context in the other part. As the contexts are connected and intersect in the splitting nodes, only one molecule is cut while the other molecules remain unchanged. Figure 2 illustrates such a splitting for a single molecule.

Now we can define splitting/fusion grammars as used in this paper. Besides fusion and splitting rules, such a grammar provides a start hypergraph, a set of markers, and a set of terminal labels. The derivation process combines rule applications with multiplications. A terminal hypergraph belongs to the generated language if it is obtained by removing all marked hyperedges from a molecule that has at least one marked hyperedge and that is derived from the start hypergraph. The markers allow one to partition the molecules of the start hypergraph into marked and unmarked ones. As markers can not be generated, the unmarked molecules can only contribute to the generated language if they are fused with marked molecules so that they are of an auxiliary nature.

Definition 4. 1. A *splitting/fusion grammar* is a system $SFG = (Z, F, M, T, SR)$ where Z is a *start hypergraph*, $F \subseteq \Sigma$ is a fusion alphabet, $M \subseteq \Sigma$ with $M \cap (F \cup \overline{F}) = \emptyset$ is a finite set of *markers*, $T \subseteq \Sigma$ with $T \cap (F \cup \overline{F}) = \emptyset = T \cap M$

is a finite set of *terminal labels*, and SR is a finite set of splitting rules that may have some type of context conditions.

2. A *direct derivation* $H \Longrightarrow H'$ for some $H, H' \in \mathcal{H}_\Sigma$ is either a rule application $H \underset{r}{\Longrightarrow} H'$ for some rule in $SR \cup \{fr(a) \mid a \in F\}$ or a multiplication $H \underset{m}{\Longrightarrow} m \cdot H$ for some multiplicity m.

3. A *derivation* $H \overset{n}{\Longrightarrow} H'$ of length n is a sequence $H_0 \Longrightarrow H_1 \Longrightarrow \cdots \Longrightarrow H_n$ with $H = H_0$ and $H' = H_n$. One may write $H \overset{*}{\Longrightarrow} H'$.

4. $L(SFG) = \{rem_M(Y) \mid Z \overset{*}{\Longrightarrow} H, Y \in \mathcal{M}(H) \cap (\mathcal{H}_{T \cup M} - \mathcal{H}_T)\}$ is the *generated language* of SFG where $rem_M(Y)$ is the hypergraph obtained when removing all hyperedges with labels in M from Y.

Remark 4. 1. A splitting/fusion grammar where $SR = \emptyset$ is called a *fusion grammar* and SR can be omitted in the tuple.

2. The provision of markers is significant. In other generative devices initial objects to start a generation and rules to perform the generation are separated while in fusion grammars the corresponding information is integrated in the start hypergraph. But it is necessary in some cases to distinguish between marked connected components that can contribute directly to the generated language and other components that can contribute to the generated languages by fusion with marked components only (cf., for example, the transformation of hyperedge replacement grammars into fusion grammars in [2]).

4   Adleman's Experiment

In this section, we adapt Adleman's famous experiment finding Hamiltonian paths by DNA computing to fusion grammars.

A graph G with designated nodes *start* and *end* is said to have a Hamiltonian path if and only if there exists a path from *start* to *end* that enters every node of the graph exactly once. The Hamiltonian path problem asks if a graph has a Hamiltonian path and has been proven to be NP-complete [6,7]. Adleman has proposed a transformation of the Hamiltonian path problem into a molecular biological process by encoding the graph by DNA molecules in order to generate a solution by massive parallelism. The computation is performed by standard DNA computing operations. In more detail, Adleman generated random paths by encoding every node of the input graph into a DNA strand such that two strands can be fused if and only if their corresponding nodes are connected via an edge in the respective direction. Afterwards filtering operations are applied to get rid of DNA strands that do not represent Hamiltonian paths from *start* to *end*. Finally emptiness is tested (see [1] for details).

For every unlabeled and loop-free graph, a fusion grammar can be constructed that generates paths from *start* to *end* by fusing smaller paths in parallel. Subsequent filter operations let only Hamiltonian paths remain. The start hypergraph of the fusion grammar contains a molecule for each edge of the input graph. Additionally, there is a molecule with a marker for constructing paths that begin with *start* and there is a molecule for the *end* node that allows to terminate paths fusions at *end*. The grammar is defined as follows.

Definition 5. Let $G = (V, E, att, lab)$ be an unlabeled simple graph, and let $start, end \in V$ with $start \neq end$. Then $FG(G) = (Z_G, \{out_v \mid v \in V\}, \{\mu\}, V \cup \{*\})$ is the fusion grammar of G, where

$$Z_G = \mu \includegraphics{start} \boxed{out_{start}} + \boxed{\overline{out_{end}}} \bullet + \sum_{e \in E:\ att(e)=vv'} \boxed{\overline{out_v}} \bullet \longrightarrow \bullet \includegraphics{v'} \boxed{out_{v'}}.$$

Example 1. Consider Adleman's sample graph, where bidirectional arrows represent two edges in opposite directions.

Select 0 as *start* and 6 as *end*. Then the start graph consists of the following molecules:

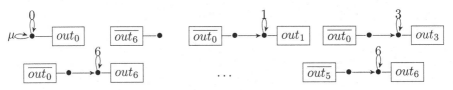

For example, the path starting from 0 and visiting then nodes 1 to 6 in this order can be sequentially generated as follows.

$$\mu \bullet \boxed{out_0} \underset{fr(out_0)}{\Longrightarrow} \mu \bullet \to \bullet \boxed{out_1} \underset{fr(out_1)}{\Longrightarrow} \mu \bullet \to \bullet \to \bullet \boxed{out_2} \underset{fr(out_2)}{\Longrightarrow} \cdots$$

$$\cdots \underset{fr(out_5)}{\Longrightarrow} \mu \bullet \to \bullet \to \bullet \to \bullet \to \bullet \to \bullet \boxed{out_6}$$

$$\underset{fr(out_6)}{\Longrightarrow} \mu \bullet \to \bullet \to \bullet \to \bullet \to \bullet \to \bullet$$

All involved fusions can be performed in parallel, because only pairwise distinct edges are fused where each such path appears with some probability depending on the number of copies of molecules of the start graph produced by a respective multiplication before the fusion step. Hence, this path (as well as all other paths) can be generated in a single step.

For each path $p = v_0 \cdots v_n$ we call the graph $pg(p) = \overset{v_0}{\bullet} \to \overset{v_1}{\bullet} \to \cdots \to \overset{v_{n-1}}{\bullet} \to \overset{v_n}{\bullet}$ a *path graph* of p.

Theorem 1. $L(FG(G)) = \{pg(p) \mid p = start\ v_1 \cdots v_n\ end$ is a path in $G\}$.

In other words, the terminal graphs generated by $FG(G)$ are all path graphs from $start$ to end.

The theorem follows directly from the following Lemma which can be proven by induction on the lengths of derivations.

Lemma 1. Let $Z_G \overset{*}{\Longrightarrow} H$ be a derivation in $FG(G)$. Let $M \in \mathcal{M}(H)$. Then M is either a molecule of Z_G or it has one of the following forms:

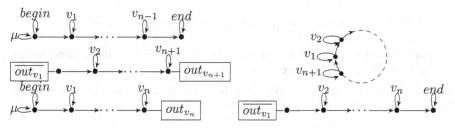

where $v_1, \ldots, v_{n+1} \in V, n \geq 1$.

In order to get only Hamiltonian paths, further filters are needed where the number of elements of a finite set X is denoted by $|X|$.

1. $length_k(L) = \{s \in L \mid |V_s| - 1 = k\}$ for $L \subseteq L(FG(G))$,
2. $simple(L) = \{s \in L \mid |V_s| = |lab_G(E_G)| - 1\}$ for $L \subseteq L(FG(G))$ where $lab_G(E) = \{lab_G(e) \mid e \in E\}$.

Corollary 1. The language $simple(length_{|V|-1}(L(FG(G))))$ is equal to the set of all path graphs $pg(p)$, where p is Hamiltonian.

The corollary indicates that it may be meaningsful to employ more general mechanisms to filter the members of the generated language from the derived hypergraphs. In the definition in Sect. 3, the generated language contains the terminal subhypergraph of a connected component of a derived hypergraph if it has some marked hyperedge, but no fusion hyperedges. This yields for our grammar that simulates Adleman's experiment graph representations of all paths from $begin$ to end so that further filtering is needed to get the Hamiltonian paths. In general, other and further filter mechanisms may be employed if it is reasonable for a specific application.

5   Transformation of Insertion-Deletion Systems into Splitting/Fusion Grammars

Insertion-deletion systems are (string) language generating devices the basic operations of which are closely related to DNA computing (see, e.g., [8] and [4]). In this section, we transform insertion-deletion systems into splitting/fusion grammars. Our main result states that the transformation is correct meaning that the language generated by an insertion-deletion system equals the language of the corresponding splitting/fusion grammar up to the representation of strings by graphs.

5.1   Insertion-Deletion Systems

An *insertion-deletion system* is a quadruple $\gamma = (V, T, A, R)$ where V is a finite alphabet, $T \subseteq V$ is a subalphabet of *terminal symbols*, $A \subseteq V^*$ is a finite language of *axioms*, and R is a finite set of *rules* of the form $r = (u, \alpha/\beta, v)$ with $u, \alpha, \beta, v \in V^*$ such that either $\alpha = \lambda$ or $\beta = \lambda$. The application of rules to strings defines a binary relation of *computation steps*:

$$insertion : w = xuvy \underset{r}{\to} xu\beta vy \text{ for } w, x, y \in V^* \text{ and } r = (u, \lambda/\beta, v), \text{ and}$$

$$deletion : w = xu\alpha vy \underset{r}{\to} xuvy \text{ for } w, x, y \in V^* \text{ and } r = (u, \alpha/\lambda, v).$$

The reflexive and transitive closure of all computation steps is called *computation relation* and denoted by $\underset{R}{\overset{*}{\to}}$, its elements are the *computations of* γ. The *generated language of* γ consists of all terminal strings that can be computed from axioms: $L(\gamma) = \{w \in T^* \mid z \underset{R}{\overset{*}{\to}} w, z \in A\}$.

5.2   String Graphs

A string is represented by a simple path where the sequence of labels along the path equals the given string.

Let Σ be a label alphabet the elements of which are of type 2. Let $w = x_1 \ldots x_n \in \Sigma^*$ for $n \geq 1$ and $x_i \in \Sigma$ for $i = 1, \ldots, n$. Then the *string graph* of w is defined by $sg(w) = (\{0\} \cup [n], [n], att_w, lab_w)$ with $att_w(i) = (i-1)i$ and $lab(i) = x_i$ for $i = 1, \ldots, n$. The string graph of λ, denoted by $sg(\lambda)$, is the discrete graph with a single node 0. Obviously, there is a one-to-one correspondence between Σ^* and $sg(\Sigma^*) = \{sg(w) \mid w \in \Sigma^*\}$.

Note that $u \in \Sigma^*$ is a substring of $w \in \Sigma^*$, i.e., $w = xuy$ for some $x, y \in \Sigma^*$ if and only if there is a graph morphism $_x u_y : sg(u) \to sg(w)$.

Each string graph $sg(w)$ for $w \in \Sigma^*$ gives rise to a special graph morphism $b_w : [2] \to sg(w)$ with $b_w(1) = 0$ and $b_w(2) = n$, where n is the length of w.

For technical reasons, we need the extension of a string graph by a labeled edge bending from the begin node to the end node. Consider $sg(w)$ for some $w \in \Sigma^*$, and let $s \in \Sigma$. Then the *s-handled string graph* $sg(w)_s$ contains $sg(w)$ as subgraph and the edge 0 with the attachment $0n$ and the label s, where n is the length of w.

5.3   The Transformation

Let $\gamma = (V, T, A, R)$ be an insertion-deletion system. The corresponding splitting/fusion grammar $SFG(\gamma)$ has a splitting rule for each rule of the insertion-deletion system, where the context u, v of the rule $(u, \alpha/\beta, v)$ is reflected by the context of the splitting rule. In the case of a deletion rule, the corresponding rule has a fixed disjoint subcontext reflecting the string to be deleted. In the case of an insertion rule, the corresponding rule has a fixed disjoint subcontext, too. But its application includes an additional node stretching within the intermediate hypergraph I'. Let $r = (u, \lambda/\beta, v)$ be an insertion rule and let $a \in \Sigma$. Then

$(sr(a), b_\lambda, {}_u\lambda_v)$ is a *splitting rule with node stretching*, and, for each $w = xuvy$, its application to $sg(w)_\mu$ with the matching $sg(uv) \xrightarrow{x\,uvy} sg(w) \subseteq sg(w)_\mu$ yields the molecules $sg(xuavy)_\mu$ and $sg(\lambda)_{\widehat{a}}$ meaning that the node separating $sg(u)$ and $sg(v)$ is stretched to an a-edge. Moreover, there is a fusion rule for each rule of the insertion-deletion system.

Definition 6. $SFG(\gamma) = (Z_\gamma, R, \{\mu\}, T, P_\gamma)$, where

- the rule set R is used as fusion alphabet with the complementary fusion alphabet $\overline{R}$ and the complementary splitting alphabet $\widehat{R}$ chosen such that $V, R, \overline{R}$ and $\widehat{R}$ are pairwise disjoint, the marker μ is an extra label, i.e., $\mu \notin V \cup R \cup \overline{R} \cup \widehat{R}$, μ and the fusion labels in R are of type 2,
- $Z_\gamma = \sum\limits_{z \in A} sg(z)_\mu + \sum\limits_{r=(u,\alpha/\beta,v)\in R} sg(\beta)_{\overline{r}}$, i.e., the start graph consists of the μ-handled string graphs of the axioms, the $\overline{r}$-handled string graphs of λ for each deletion rule r, and the $\overline{r}$-handled string graphs of the insertion string of each insertion rule r, and
- P_γ contains two types of splitting rules:
  1. for each deletion rule $r = (u, \alpha/\lambda, v) \in R$, there is a splitting rule with fixed disjoint subcontext $r_\gamma = (sr(r), b_\alpha, {}_u\alpha_v)$,
  2. for each insertion rule $r = (u, \lambda/\beta, v) \in R$, there is a splitting rule with node stretching $r_\gamma = (sr(r), b_\lambda, {}_u\lambda_v)$.

To see how these rules work, consider $sg(w)_\mu$. The rule r_γ for $r = (u, \alpha/\lambda, v)$ is applicable if $w = xu\alpha vy$ for some x, y using the matching $sg(u\alpha v) \xrightarrow{x\,u\alpha vy} sg(w) \subseteq sg(w)_\mu$. Then the splitting produces the two molecules $sg(xurvy)_\mu$ and $sg(\alpha)_{\widehat{r}}$. If the former molecule is fused with $sg(\lambda)_{\overline{r}}$ applying the fusion rule $fr(r)$, one gets the molecule $sg(xuvy)_\mu$. In other words, the application of r_γ to $sg(w)_\mu$ followed by the application of $fr(r)$ coincides with the computation step $w \xrightarrow{r} xuvy$ in γ up to representation of strings by graphs. Similarly, the rule r_γ for $r = (u, \lambda/\beta, v)$ is applicable to $sg(w)_\mu$ if $w = xuvy$ for some x, y using the matching $sg(uv) \xrightarrow{x\,u\lambda v y} sg(w) \subseteq sg(w)_\mu$. The splitting with stretching yields the molecules $sg(xurvy)_\mu$ and $sg(\lambda)_{\widehat{r}}$. If the former molecule is fused with $sg(\beta)_{\overline{r}}$ applying the fusion rule $fr(r)$, then one gets $sg(xy\beta vy)_\mu$. This means that the application of r_γ to $sg(w)_\mu$ followed by the application of $fr(r)$ coincides with the computation step $w \xrightarrow{r} xu\beta uv$ in γ.

These observations allow to prove the following lemma that links computations in an insertion-deletion system to special derivations in the corresponding splitting/fusion grammar. To formulate the lemma, we use three kinds of multiplication for a given insertion-deletion system $\gamma = (V, T, A, R)$:

1. $m(z)$ for $z \in A$ removes all molecules $sg(z')_\mu$ (via multiplication by 0) for $z' \in A$ with $z' \neq z$ and keeps all others.
2. $m(r)$ for $r = (u, \alpha/\beta, v) \in R$ duplicates the molecule $sg(\beta)_{\overline{r}}$ and keeps all others.
3. $m(\widehat{r})$ for $r = (u, \alpha/\beta, v) \in R$ removes the molecule $sg(\alpha)_{\widehat{r}}$ and keeps all others.

Lemma 2. Let $\gamma = (V, T, A, R)$ be an insertion-deletion system and $SFG(\gamma)$ the corresponding splitting/fusion grammar. Let $d = (w_0 \underset{r_1}{\to} w_1 \underset{r_2}{\to} \ldots \underset{r_n}{\to} w_n)$ be a computation in γ with $w_0 \in A$ and $r_i = (u_i, \alpha_i/\beta_i, v_i) \in R$. Then there is a derivation in $SFG(\gamma)$ of the following form:

$$d_\gamma = (Z_\gamma \underset{m(w_0)}{\Longrightarrow} sg(w_0)_\mu + X_\gamma \overset{4}{\Longrightarrow} sg(w_1)_\mu + X_\gamma \overset{4}{\Longrightarrow} \cdots \overset{4}{\Longrightarrow} sg(w_n)_\mu + X_\gamma),$$

where $X_\gamma = \underset{r=(u,\alpha/\beta,v)\in R}{\sum} sg(\beta)_{\overline{r}}$. The sections $sg(w_{i-1})_\mu + X_\gamma \overset{4}{\Longrightarrow} sg(w_i)_\mu + X_\gamma$ for $i = 1, \ldots, n$ are defined by

$$sg(w_{i-1})_\mu + X_\gamma \underset{(r_i)_\gamma}{\Longrightarrow} sg(x_i u_i r_i v_i y_i)_\mu + X_\gamma + sg(\alpha_i)_{\widehat{r}_i} \underset{m(\widehat{r}_i)}{\Longrightarrow} sg(x_i u_i r_i v_i y_i)_\mu + X_\gamma$$

$$\underset{m(r_i)}{\Longrightarrow} sg(x_i u_i r_i v_i y_i)_\mu + sg(\beta_i)_{\overline{r}_i} + X_\gamma \underset{fr(r_i)}{\Longrightarrow} sg(w_i)_\mu + X_\gamma$$

for some $x_i, y_i \in (V \cup R)^*$.

The derivation d_γ is called *insdel*-derivation of d.

Conversely, the derivations in $SFG(\gamma)$ can also be nicely related to the computations in γ.

Lemma 3. Let $\gamma = (V, T, A, R)$ be an insertion-deletion system and $SFG(\gamma)$ the corresponding splitting/fusion grammar. Let $D = (Z_\gamma \overset{*}{\Longrightarrow} H)$ be a derivation in $SFG(\gamma)$ and $sg(w)_\mu \in \mathcal{M}(H)$ for some $w \in V^*$. Then there is an insdel-derivation d_γ of some computation $d = (w_0 \overset{*}{\to} w)$.

Using these lemmata, one can prove that an insertion-deletion system and the corresponding splitting/fusion grammar generate the same language up to the representation of strings as string graphs.

Theorem 2. *Let $\gamma = (V, T, A, R)$ be an insertion/deletion system and $SFG(\gamma)$ its corresponding splitting/fusion grammar. Then*

$$sg(L(\gamma)) = \{sg(w) \mid w \in L(\gamma)\} = L(SFG(\gamma)).$$

6   2-Splicing Grammars

In the literature, one encounters many variants of splicing systems as they are considered as a potential computational kernel of a future DNA computer (see e.g., [4] for a comprehensive survey). Typically, a rule of a splicing system has the form of a quadruple $(u_1, u_2; u_3, u_4)$ of four strings. It is applicable to two strings w and w' if $w = x_1 u_1 u_2 x_2$ and $w' = x_3 u_3 u_4 x_4$ for some strings x_1, x_2, x_3, and x_4. Such a rule application splits w and w' between u_1, u_2 and u_3, u_4 respectively and recombines the parts into $x_1 u_1 u_4 x_4$ and $x_2 u_2 u_3 x_3$. The operation is an 1-splicing if only the first result is further taken into account; it is a 2-splicing if both results are further considered. In an iterated splicing system, the splicing process is arbitrarily iterated on the resulting strings starting with a given set of strings, called axioms. Finally, an extended iterated splicing system has an additional alphabet of terminal symbols, and its generated language consists of all terminal strings that result from the splicing process. In this section, we introduce 2-splicing grammars generalizing extended iterated 2-splicing systems to hypergraphs as underlying structures.

Definition 7. 1. A *2-splicing grammar* is a system $2SG = (V, T, A, R)$ where V is a finite label alphabet, $T \subseteq V$ is a subalphabet of *terminal labels*, $A \subseteq \mathcal{H}_V$ is a finite set of connected hypergraphs, called *axioms*, and R is a finite set of *rules* of the form $r = (c_1, c_2; c_3, c_4)$ with $c_i \colon [k(r)] \rightarrow C_i$ where C_i is a connected hypergraph for each $i = 1, 2, 3, 4$ and some $k(r) \in \mathbb{N}$, called *type of* r.

2. The application of r to $H \in \mathcal{H}_V$ is defined by the application of the two splitting rules with double context $(sr(r_1), c_1, c_2)$ and $(sr(r_2), c_3, c_4)$ to two different molecules of H followed by the application of the fusion rules $fr(r_1)$ and $fr(r_2)$ where the complementary fusion and splitting labels satisfy the condition $\overline{r}_1 = \widehat{r}_2$ and $\overline{r}_2 = \widehat{r}_1$.

3. A *derivation* in $2SG$ from H to H' is a sequence of rule applications and multiplications $H = H_0 \Longrightarrow H_1 \Longrightarrow \ldots \Longrightarrow H_n = H'$, shortly denoted by $H \overset{*}{\Longrightarrow} H'$.

4. The *generated language* of $2SG$ consists of all terminal molecules of hypergraphs derived from the axioms: $L(2SG) = \{Y \in \mathcal{H}_T \mid \sum_{Z \in A} Z \overset{*}{\Longrightarrow} H, Y \in \mathcal{M}(H)\}$.

Remark 5. 1. Focusing on rule application to the two changed molecules, it looks as in Fig. 3 where the subgraphs $f(C_j)$, $f'(C_j)$ and $f''(C_j)$ for $j = 1, 2$ coincide up to the distinguished nodes.

2. Without loss of generality, one can assume that none of the molecules involved in a derivation disappears because one may duplicate the two molecules that are cut by the rule application beforehand.

3. Note that the 2-splicing grammars are defined without markers. We refrain from their use because all the molecules of the start hypergraph can contribute to the generated language in the same way.

Example 2. To illustrate the concept of 2-splicing grammars, we specify a sample grammar MOP that generates a certain type of maximal outerplanar graphs. An undirected unlabeled graph is a *maximal outerplanar graph* (*mop* for short) if it consists of a simple cycle that visits all nodes and a maximum number of further edges such that the graph is planar, but any further edge would yield a non-planar graph.

$$MOP = (\{*\}, \{*\}, \{\triangle, \bigtriangledown\!\!\!\!\bigtriangleup\}, \{r_M = (\prec_{2}^{1}, \triangleright; \prec_{2}^{1}, \triangleright)\})$$

where the type of the single rule is 2 and the morphisms from the discrete graph with two nodes to the contexts are indicated by the numbered nodes.

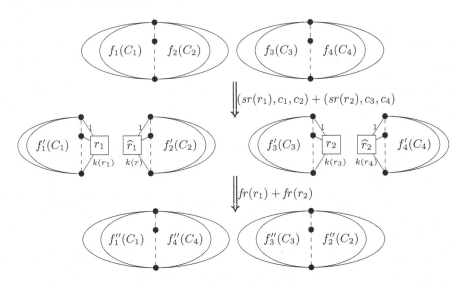

Fig. 3. 2-splicing wrt two molecules

The start graph joins the mops with 3 and 5 nodes. The splitting rule with the first two graphs of the rule as double context cannot be applied to the triangle as the embedding of the two context graphs are required to intersect in the distinguished nodes only. But the rule of MOP can be applied to two copies of the mop with 5 nodes in the following way:

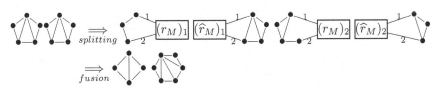

An alternative way of the splitting yields

Then the recombination of the third and the second molecule yields .

This means that rule application in MOP can generate the mop with 4 nodes and two mops with 6 nodes. Iterating the rule application, one can generarate mops with an arbitrary number of nodes.

Let $2SG_{sg} = (V, T, A, R)$ be a 2-splicing grammar such that $A \subseteq sg(V^*)$ and each rule $r \in R$ has the form $(b_{u_1}, b_{u_2}; b_{u_3}, b_{u_4})$ for some $u_1, u_2, u_3, u_4 \in V^*$ (cf. Sect. 5.2). Then $\sigma = (V, T, A_\sigma, R_\sigma)$ with $A_\sigma = \{z \in V^* \mid sg(z) \in A\}$ and

$R_\sigma = \{(u_1, u_2; u_3, u_4) \mid (b_{u_1}, b_{u_2}; b_{u_3}, b_{u_4}) \in R\}$ defines an extended iterated 2-splicing system (on strings) that is also called extended H-system in [4]. This means that there is an obvious relation between the string case and the hypergraph case on the syntactic level. In contrast to this, the generated language of an extended iterated 2-splicing system is defined by means of an infinite iteration on sets of strings and not by derivations. The iteration process can be carried over to the hypergraph case.

Definition 8. Let $2SG = (V, T, A, R)$ be a 2-splicing grammar. Then its *iterated language* $L'(2SG)$ is defined as follows: $L'(2SG) = \sigma_2^*(A) \cap T^*$ with $\sigma_2^*(A) = \bigcup_{i \in \mathbb{N}} \sigma_2^i(A)$ where $\sigma_2^i(A)$ is inductively defined by $\sigma_2^0(A) = A$, and $\sigma_2^{i+1}(A) = \sigma_2(\sigma_2^i(A))$ for all $i \in \mathbb{N}$ with $\sigma_2(X) = \{M_3, M_4 \mid M_1 + M_2 \underset{r}{\Longrightarrow} M_3 + M_4, r \in R, M_1, M_2 \in X\}$ for all sets X of connected hypergraphs in $\mathcal{H}_V$.

Nicely enough, it turns out that the generated language and the iterated language of a 2-splicing grammar coincide.

Theorem 3. *Let $2SG$ be a 2-splicing grammar. Then $L(2SG) = L'(2SG)$.*

This means that the hypergraph case is a proper generalization of the string case on the semantic level, too.

7   Conclusion

In this paper, we have related three well-known DNA computing approaches to the framework of splitting/fusion grammars. First, we have recreated Adleman's seminal experiment that marks the origin of DNA computing by means of fusion grammars (where no splitting is needed). Secondly, we have transformed insertion-deletion systems into splitting/fusion grammars. And, thirdly, we have generalized extended iterated 2-splicing systems, that process strings as underlying data structure, by 2-splicing grammars that operate on hypergraphs. Future research in this context may head in various directions:

Further approaches to DNA computing like sticker systems (cf., e.g., [4]) may be related to splitting/fusion grammars.

In analogy to Adleman's experiment, we have solved the Hamiltonian path problem by fusion grammars with additional filter mechanisms in such a way that the decision takes one multiplication step and one parallel fusion step and is correct with high probability. It may be interesting to consider other NP-complete problems and to investigate the computational capability of splitting/fusion grammars in more depth.

In addition to the transformation of insertion-deletion systems, it may be worthwhile to generalize this kind of string processing to hypergraph processing with the hope that interesting examples can be modeled in this way.

Besides extended iterated 2-splicing systems, one encounters many variants based on splicing in the literature. To consider them from the point of view of splitting/fusion grammars may lead to new insights.

In [9], graph multiset transformation was introduced as a computational approach with massive parallelism inspired by DNA computing like splitting/fusion grammars. There the traditional double-pushout rules are used rather than the very special cases of splitting and fusion rules. As multisets and disjoint unions are very similar data structures, a comparison of the two approaches may be worthwhile.

Acknowledgment. We are grateful to the anonymous reviewers for their critical comments that encouraged us to add some more explanations.

References

1. Adleman, L.M.: Molecular computation of solutions to combinatorial problems. Science **266**, 1021–1024 (1994)
2. Kreowski, H.-J., Kuske, S., Lye, A.: Fusion grammars: A novel approach to the generation of graph languages. In: de Lara, J., Plump, D. (eds.) ICGT 2017. LNCS, vol. 10373, pp. 90–105. Springer, Cham (2017). https://doi.org/10.1007/978-3-319-61470-0_6
3. Kreowski, H.-J., Kuske, S., Lye, A.: Splicing/fusion grammars and their relation to hypergraph grammars. In: Lambers, L., Weber, J. (eds.) ICGT 2018. LNCS, vol. 10887, pp. 3–19. Springer, Cham (2018). https://doi.org/10.1007/978-3-319-92991-0_1
4. Păun, G., Rozenberg, G., Salomaa, A.: DNA Computing—New Computing Paradigms. Springer, Heidelberg (1998)
5. Kreowski, H.-J., Klempien-Hinrichs, R., Kuske, S.: Some essentials of graph transformation. In: Ésik, Z., Martín-Vide, C., Mitrana, V. (eds.) Recent Advances in Formal Languages and Applications. Studies in Computational Intelligence, vol. 25, pp. 229–254. Springer, Heidelberg (2006). https://doi.org/10.1007/978-3-540-33461-3_9
6. Garey, M.R., Johnson, D.S.: Computers and Intractability: A Guide to the Theory of NP-Completeness. W. H. Freeman, London (1979)
7. Karp, R.M.: Reducibility among combinatorial problems. In: Miller, R.E., Thatcher, J.W., (eds.) Proceedings of a Symposium on the Complexity of Computer Computations, The IBM Research Symposia Series, pp. 85–103. Plenum Press, New York (1972)
8. Kari, L., Thierrin, G.: Contextual insertions/deletions and computability. Inf. Comput. **131**(1), 47–61 (1996)
9. Kreowski, H.-J., Kuske, S.: Graph multiset transformation as a framework for massively parallel computation. In: Ehrig, H., Heckel, R., Rozenberg, G., Taentzer, G. (eds.) ICGT 2008. LNCS, vol. 5214, pp. 351–365. Springer, Heidelberg (2008). https://doi.org/10.1007/978-3-540-87405-8_24

Transformation Rules Construction and Matching

Constructing Optimized Validity-Preserving Application Conditions for Graph Transformation Rules

Nebras Nassar$^{(\boxtimes)}$ (ID), Jens Kosiol (ID), Thorsten Arendt (ID),
and Gabriele Taentzer (ID)

Philipps-Universität Marburg, Marburg, Germany
{nassarn,kosiolje,taentzer}@informatik.uni-marburg.de,
thorsten.arendt@uni-marburg.de

Abstract. There is an increasing need for graph transformations ensuring valid result graphs wrt. a given set of constraints. In a model refactoring process, for example, each performed refactoring should yield a valid model graph. At least, it has to remain an element of the underlying modeling language. If a graph transformation rule always produces valid output, it is called *validity-guaranteeing*; if only when applied to an already valid graph, it is called *validity-preserving*. There is a formal construction for graph transformation systems making them validity-guaranteeing. This is ensured by adding a validity-guaranteeing application condition to each of its transformation rules. This theory has been implemented recently as an Eclipse plug-in called OCL2AC. Initial tests have shown that resulting application conditions can become pretty large. As there are interesting application cases where transformations just need to be validity-preserving (such as model refactoring), we started to investigate this case further. The results are optimizing-by-construction techniques for application conditions for transformations that just need to be validity-preserving. All presented optimizations are proven to be correct. Implementing and evaluating them, we found that the complexity of the resulting application conditions is considerably reduced (by factor 7 on average). Moreover, our optimization yields a speedup of rule application by approximately 2.5 times.

Keywords: Graph transformation · Constraints · Correctness

1 Introduction

Model transformations are the heart and soul of Model-Driven Engineering (MDE). They are used for various MDE-activities including translation, optimization, and synchronization of models [31]. Usually, a transformation (that may consist of several transformation steps) should yield a valid result model, especially if it has been applied to an already valid model. Intermediate models

© Springer Nature Switzerland AG 2019
E. Guerra and F. Orejas (Eds.): ICGT 2019, LNCS 11629, pp. 177–194, 2019.
https://doi.org/10.1007/978-3-030-23611-3_11

may not be required to be valid as, e.g., argued in [8]. But there are scenarios where even intermediate models have to show validity, at least a basic one, as the following example applications show: (1) Throughout a larger refactoring process, each performed refactoring should preserve the model's validity [3]. (2) More generally, any in-place model change should preserve a basic validity, enough to view an edited model in its domain-specific model editor [16]. Model editors typically ensure the creation of models with basic validity right from the beginning. This is the application scenario we will use as running example and for our evaluation. A similar scenario is considered in projectional editing for textual editors [32]. (3) Modeling the behavior of concurrent and distributed systems with model transformations, each model represents a system state that should fulfill system invariants such as safety properties [17]. (4) When generating code from abstractly specified model transformations, the transformations should be validity-preserving, especially for safety-critical systems [11].

State of the Art. From the formal point of view, the theory of algebraic graph transformation constitutes a suitable framework to reason about model transformations [9,10], in particular about rule-based transformation of EMF models [4]. Constraints are typically expressed as (nested) graph constraints [13,29], into which a large and relevant part of OCL [24] can be translated [28]. Graph constraints can be integrated as application conditions into graph transformation rules as shown in [13]. Given a rule and a constraint, there are two variants of integration, namely computing a *constraint-preserving* or a *constraint-guaranteeing* rule. Both computations do not alter the actions of the rule but equip it with an application condition. Graph validity is *preserved*, if applying an equipped rule to a valid graph, the resulting graph is valid as well. Graph validity is *guaranteed*, if applying an equipped rule to a graph, the resulting graph is valid. As for tool support, OCL2AC [19] automatically translates OCL constraints into graph constraints and integrates these as application conditions into transformation rules specified in Henshin [1]. It computes guaranteeing rules.

Tests of OCL2AC have shown that resulting application conditions can become very complex. Theoretically, application conditions of guaranteeing rules grow over-exponentially in the worst case [26]. As there are interesting application cases where transformations just need to be validity-preserving (as pointed out above), it is worthwhile to investigate validity-preserving transformations further. Habel and Pennemann [13] present a direct construction of the logically weakest application condition, enough to preserve validity. As this kind of condition is logically weaker, our expectation was in the beginning that it can be expressed in a simpler form. In contrast, the resulting application conditions may contain even more elements than the validity-guaranteeing ones. This is due to the approach taken: The premise that the model was already valid before rule application is added to the computed validity-guaranteeing application condition. The resulting condition can be inherently difficult to simplify because of the used material implication operator. An example is presented in [20].

Contribution and Structure. Focusing on validity preserving transformations only, we develop optimizing-by-construction techniques to construct application conditions that preserve validity and are considerably less complex than the results of the original construction.

1. In Sect. 4, we take a constraint and a rule as starting point and construct an application condition that preserves validity. This construction is based on the construction of the guaranteeing application conditions but simplifies it by omitting parts that check for antecedent validity, while keeping parts that prevent the introduction of violations. This automatic *approximation* of the preserving application condition is conceptually new and quite general in scope. While some of the simplifications are specific for EMF (Theorem 2), the others (Theorem 1) are proven for graph constraints in general and can be easily lifted to adhesive categories [18]. We will argue how some of these simplifications omit *global* checks that have to traverse the whole model while keeping *local* ones, i.e., checks being performed in the context of a rule match.
2. Practically, we have implemented the techniques on top of OCL2AC (Sect. 5) and compared the application conditions of guaranteeing rules with those of preserving ones. The results show a considerable loss in complexity of application conditions (Sect. 6.1).
3. We provide an application case which shows that validity-preserving transformations are useful in practice. In domain-specific model editing (presented as scenario (2) above), every state of the transformation process has to ensure a basic model validity. The example comprises the MagicDraw Statechart meta-model with 11 OCL constraints and 84 editing rules. The optimizations do not only reduce the size of computed application conditions considerably but also improve the performance of validity-preserving transformations.
   In addition, we have conducted several evaluations that do not specifically test our optimization but the overall approach. We compared the run times of validity checking after a transformation using existing OCL validators (*a posterori approach*) with running a validity-preserving transformation (being enriched with application conditions) with and without optimization (*a priori approach*) (Sect. 6.2). Results show that both approaches are fast in practice. Actually, it is the first time that the usability of OCL2AC, and the implemented approach in general, is investigated.

We start our presentation with the running example in Sect. 2 and recall the formal and technical background in Sect. 3. All proofs and more details about the evaluation can be found in an extended version of the paper [20].

2   Running Example

In this section, we illustrate the effect of our optimizations on application conditions computed by OCL2AC.

A simple Statecharts language serves as an example. Its meta-model is displayed in Fig. 1. A StateMachine contains at least one Region and Pseudostates

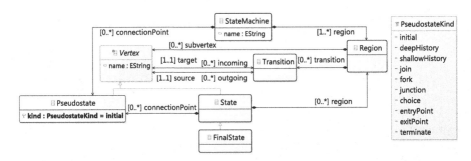

Fig. 1. A simple Statecharts meta-model

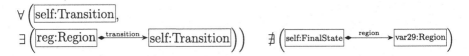

Fig. 2. Graph constraint for TransitionIn-Region

Fig. 3. Graph constraint for no_region

as connection points if they are of kind entryPoint or exitPoint. A Region contains Transitions and Vertices. Vertex is an abstract class with concrete subclasses State and Pseudostate. A State may contain Regions and Pseudostates to support the specification of state hierarchies. FinalState inherits from State. Transitions connect Vertices.

The UML definition specifies several constraints on statechart models. For example, each Transition is required to be contained in a Region (TransitionInRegion) and a FinalState is forbidden to contain a Region (no_Region). Figures 2 and 3 show these constraints as graph constraints, respectively. In the UML, however, these constraints are specified in OCL; the OCL constraint for no_region, for example, is specified as

context FinalState **invariant** no_region: self.region–>isEmpty()

Figure 4 shows a simple transformation rule in Henshin taken from [16] for specifying an edit operation in MagicDraw [21]. The rule moves an existing Region from an existing State (the old source) to another existing State (the new source). This is done by deleting the containment edge region from the old source and recreating it in the new source. Rules specifying such

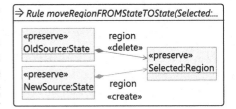

Fig. 4. Transformation rule in Henshin

edit operations may be used, e.g., to recognize semantic change sets while comparing two model versions [15,16].

The validity of basic constraints should be preserved throughout editing because a typical model editor is not able to display an instance violating them. Since FinalState is a subtype of State, applying the rule moveRegionFromStateToState might introduce a violation of the constraint no_region. Using OCL2AC [19], a language engineer can automatically integrate a constraint as an application condition into the rule and calculate the according constraint-guaranteeing version of the rule. The guaranteeing application condition obtained by integrating constraint no_region into rule moveRegionFromState-ToState forbids matching this rule to a FinalState. It checks additionally if the model already encompasses a FinalState containing a Region – either matched by the rule or not. Figure 5 presents the resulting guaranteeing application condition which is composed of 7 graphs (explained later in Sect. 4.1). Knowing the input model to be valid, most of the checks are unnecessary. Especially the checks which do not only involve elements being local to the rule application but amount to traversing every existing node, i.e., the global checks.

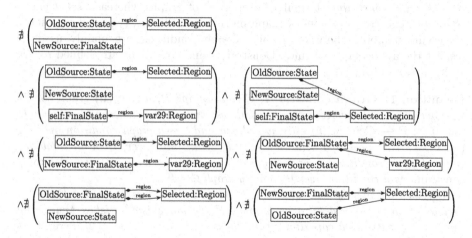

Fig. 5. Non-optimized application condition for moveRegionFromStateToState after integrating the constraint no_region

In this paper, we develop and implement optimizations that allow for omitting certain parts from the construction of a guaranteeing application condition. In our example, we will arrive at the optimized application condition shown in Fig. 6 which consists of only one graph that, moreover, only requires a local check. It forbids the rule node newSource:State to be matched to a FinalState.

As the rule moveRegionFromState-ToState does not change any graph element occurring in constraint TransitionInRegion, this constraint cannot be violated by a result model if it was not violated before. Hence, the optimized

$\nexists \left(\begin{array}{l} \boxed{\text{OldSource:State}} \xleftarrow{\text{region}} \boxed{\text{Selected:Region}} \\ \boxed{\text{NewSource:FinalState}} \end{array} \right)$

Fig. 6. Optimized application condition for moveRegionFromStateToState still preserving the constraint no_region

application condition is just true. The guaranteeing condition (not shown), however, consists of three graphs. Thus, assuming valid input models, guaranteeing application conditions can be considerably simplified.

3   Formal Background and Tooling

Our approach is based on the theory of algebraic graph transformation [9]. EMF models and model transformations are formalized as typed attributed graphs and graph transformations as presented in [4]. In the following, we recall (i) *nested graph constraints* and *conditions* as a means to express properties of graphs and graph morphisms and (ii) graph transformation rules as our formal background. Besides, we mention OCL2AC as a tool support.

3.1   Constraints, Conditions, and Rules

Nested graph constraints formulate properties of graphs whereas *nested graph conditions* express properties of graph morphisms [13], i.e., type and structure-preserving mappings between graphs. Graph conditions are mainly used to restrict the applicability of rules. Constraints and conditions are defined recursively as trees of injective morphisms.

Definition 1 (Graph condition). *Given a graph P, a (nested) graph condition over P is defined recursively as follows: true is a graph condition over P and if $a : P \hookrightarrow C$ is an injective morphism and c is a graph condition over C, $(a : P \hookrightarrow C, c)$ is a graph condition over P again. Moreover, Boolean combinations of graph conditions over P are graph conditions over P. A (nested) graph constraint is a condition over the empty graph $\varnothing$.*

Satisfaction of a graph condition d over P for a morphism $p : P \rightarrow G$, denoted as $p \models d$, is defined as follows: Every morphism satisfies true. The morphism p satisfies a condition of the form $d = \exists (a : P \hookrightarrow C, c)$ if there exists an injective morphism $q : C \hookrightarrow G$ such that $p = q \circ a$ and q satisfies c. For Boolean operators, satisfaction is defined as usual. A graph G satisfies a graph constraint d, denoted as $G \models d$, if the empty morphism to G does so.

Graph constraints are expressively equivalent to a first-order logic on graphs [13, 29]. To ease notation, we drop the domain of morphisms in constraints and conditions whenever they may be unambiguously inferred and indicate the mapping by the names of nodes. We call constraints of the form $\exists C$ *positive* and of the form $\neg \exists C$ *negative constraints*. Examples for graph constraints and conditions with informal explanation of their semantics are given in Sect. 2.

Rules are a technical means to declaratively define model transformations.

Definition 2 (Rule Transformation). *A rule $\rho = (p, lac, rac)$ consists of a plain rule p and left and right application conditions lac and rac. The plain rule p consists of three graphs $L, K,$ and R, called left-hand side (LHS), interface, and right-hand side (RHS) with two inclusion morphisms $l : K \hookrightarrow L, r : K \hookrightarrow R$.*

A rule p is monotonic *if $l : K \hookrightarrow L$ is an isomorphism and only deletes if $r : K \hookrightarrow R$ is an isomorphism. The application conditions lac and rac are graph conditions over L and R, respectively.*

Given a rule $\rho = ((L \overset{l}{\hookleftarrow} K \overset{r}{\hookrightarrow} R), lac, rac)$ and an injective morphism $m : L \hookrightarrow G$ with $m \vDash lac$, called match, *a (direct) transformation $G \Rightarrow_{\rho,m} H$ from G to H via ρ at match m is given by the diagram to the right where both squares are pushouts.*

$$
\begin{array}{ccccc}
L & \longleftarrow & K & \hookrightarrow & R \\
{\scriptstyle m \vDash lac}\downarrow & & \downarrow & & \downarrow{\scriptstyle n \vDash rac} \\
G & \longleftarrow & D & \hookrightarrow & H
\end{array}
$$

A rule p is applicable *at match m if the first pushout square above exists, i.e., if $m \circ l$ has a pushout complement D, and, moreover, the match morphism m satisfies lac and the co-match n satisfies rac.*

Note that the first pushout square exists if and only if the match m fulfills the dangling edge check ensuring that a rule application at this match would not let an edge dangle. Applying the rule, the elements of $m(L \backslash K)$ are deleted. Then, at the chosen image of K in G, a copy of $R \backslash K$ is created. Afterwards, the resulting mapping of the graph R into the new graph is checked to fulfill the right application condition of the rule. In that case, the new graph is the result of the rule application.

An example of a rule is shown in Fig. 4. Right application conditions are important in theory but not necessary in practice as they may equivalently be transformed into left application conditions. Therefore, application conditions are understood to be left application conditions.

Computing Application Conditions from Graph Constraints. Given a rule and a constraint, one computes all the different ways in which the constraint may be satisfied after applying the rule. This is done by overlapping its RHS in all possible ways with the graphs of the constraint. This computation is iterated along the nesting structure of the constraint. The result is a right application condition for the rule that is satisfied only if the constraint is valid after rule application. By applying the inverse rule to this right application condition, again along its nesting structure, a left application condition is received still *guaranteeing* validity w.r.t. the given constraint. Adding the premise that the constraint was already valid before rule application yields the *preserving application condition*.

Starting in [14] for special cases in the category of graphs, this construction has been generalized to arbitrary nested constraints in the general setting of $\mathcal{M}$-adhesive categories [13].

Fact 1 ([13]). *Given a plain rule $p = (L \hookleftarrow K \hookrightarrow R)$ and a graph constraint c there are constructions $\mathsf{Gua}(p, c)$ and $\mathsf{Pres}(p, c)$ equipping p with an application condition ac such that $H \vDash c$ for every transformation $G \Rightarrow_{\mathsf{Gua}(p,c)} H$ and $H \vDash c$ for every transformation $G \Rightarrow_{\mathsf{Pres}(p,c)} H$ where $G \vDash c$.*

3.2 OCL2AC Tool

OCL2AC [19] is an Eclipse plug-in implementing the existing theory [13,28] for adapting a given rule-based model transformation such that resulting models guarantee a given constraint set. OCL2AC consists of two main components: (1) OCL2GC takes a meta-model [7] and a set of OCL constraints as inputs and automatically returns a set of semantically equivalent graph constraints as output. (2) GC2AC takes a transformation rule defined in Henshin and a graph constraint and automatically returns the Henshin rule with an updated application condition guaranteeing the given graph constraint. Each component can be used independently as an Eclipse-based tool.

Limitations. The general formal approach we are based on, and hence OCL2AC as well, come with the following limitations: The supported logic is two-valued and first-order and thus the expression oclIsUndefined and the operation iterate are not supported, for example. Moreover, there is no support to translate user-defined operations and there is only limited support to integrate constraints on attributes into Henshin rules that perform complex attribute computations.

4   Optimizing Application Conditions

The application conditions being calculated by the approach of the tool OCL2AC guarantee validity even if the input is not a valid EMF-model. Since we focus on validity preservation of EMF-models in this paper, the calculated conditions can be considerably simplified. In this section, we investigate several strategies to construct optimized validity-preserving application conditions.

4.1 Approximating Preservation

In common application scenarios (like refactoring), a user can assume that rules are applied to instances showing a certain validity. Hence, when applying a rule, an already valid constraint does not need to be guaranteed but just preserved. The construction Pres of a preserving rule (as mentioned in Fact 1) takes this into account. Though being logically weaker, the resulting application condition can be even more complex with respect to the structure and number of contained graphs and simplification is inherently difficult. Nevertheless, it is possible to simplify guaranteeing application conditions *during the construction process* if they just need to preserve validity. In the following, we present three forms of simplification.

1. We collect all rule elements being deleted or created and check if this set overlaps with the set of all constraint elements. If this overlap is empty, the resulting preserving application condition is just true.
2. If a rule creates new graph structure only, positive constraints $\exists C$ do not need to be integrated into such a rule. Analogously, if a rule only deletes graph structure, negative constraints $\neg \exists C$ do not need to be integrated. In both cases, applications of such a rule cannot introduce a new violation of the constraint. Hence, the optimized application condition is just true.

3. When calculating an application condition, a constraint graph is overlapped with the RHS graph of a rule in all possible ways. For negative constraints $\neg \exists C$ it is not necessary to consider all possible overlappings. One may omit all the cases where C and the RHS R do not overlap in at least one element created. The parts of the application conditions arising from those cases would just check that the input graph already fulfills the constraint.

Especially the third simplification omits cases where the arising graph in the application condition contains nodes not connected to nodes of the LHS of the rule, thus amounting to global checks upon application. We state the correctness of these simplifications in the following theorem.

Theorem 1 (Correctness of simplifications). *Let c be a graph constraint and $p = (L \hookleftarrow K \hookrightarrow R)$ be a plain rule. Let $\rho = (p, ac)$ be the same plain rule equipped with the application condition ac computed in one of the following ways:*

1. *If both the elements of $L \backslash K$ and the elements of $R \backslash K$ intersect emptily with every graph C occurring in the constraint c, then $ac = \mathbf{true}$.*
2. *If p is monotonic and c is a positive constraint, then $ac = \mathbf{true}$. Analogously, if p only deletes and c is a negative constraint, then $ac = \mathbf{true}$.*
3. *If $c = \neg \exists C$, let $Gua(p, c)$ yield the right application condition $rac := \neg(\bigvee_{i \in I} \exists P_i)$ with morphisms $c_i : C \hookrightarrow P_i$ and $r_i : R \hookrightarrow P_i$. Let $rac_{pres} := \neg(\bigvee_{j \in J} \exists P_j)$ with $J \subseteq I$ including only those P_i where $c_i(C) \cap r_i(R \backslash K) \neq \varnothing$. Then ac is the application condition that arises by translating the right application condition rac_{pres} to the LHS of rule p.*

Then for all transformations $G \Rightarrow_{\rho = (p, ac)} H$ where $G \models c$ also $H \models c$.

The proof follows a common pattern in all cases: Checking for the (non-)existence of graphs occurring in the constraint in all these cases is *sequentially independent* from application of the rule. Hence, checking the constraint for validity always gives the same result, no matter if done before or after rule application.

Example 1 (compare Sect. 2). Constraint no_region is required to be integrated into rule moveRegionFromStateToState since a region-edge is created by this rule and contained in this constraint. Figure 5 shows the guaranteeing application condition. The first graph (the uppermost graph) results from a maximal overlapping of the constraint with the rule. Note that it is possible to identify nodes of types State and FinalState since FinalState is a subtype of State (compare Fig. 1). The second graph results from copying the graph of the constraint and the RHS of the rule and putting them next to each other. The third graph results from merging the nodes of type Region. The forth and the fifth graph result from just merging nodes of type State and FinalState. The sixth and the seventh graph result from merging the nodes of type State and the nodes of type Region. In every case, the overlapping of the constraint with the RHS is then translated to the LHS of the rule.

Our proposed optimizations lead to the result displayed in Fig. 6 by the application of Theorem 1, 3.: Except for the subcondition containing the uppermost graph, all other subconditions in Fig. 5 are omitted. The uppermost one has to be saved because the region-edge created by the rule is overlapped with the region-edge of the constraint. The omitted subconditions do not only involve elements being local to the rule application but amount to traversing every existing FinalState leading to global checks. To conclude, only one local check remains.

Example 2 (compare Sect. 2). The constraint TransitionInRegion is not required to be integrated into the rule moveRegionFromStateToState. Theorem 1, 1. justifies this: the rule moveRegionFromStateToState does not have any effect on the validity of the constraint since its application neither deletes nor creates elements that occur in the constraint.

4.2   Dealing with EMF's Built-in Negative Constraints

EMF has several built-in constraints [4]. Instance models that do not satisfy these EMF-constraints cannot even be opened in the EMF-editor. Most of these constraints are negative, i.e., they forbid certain patterns in instances to exist. Concretely, cycles over containment edges, nodes with more than one container, and parallel edges, i.e., two edges of the same type between the same two nodes, are forbidden. Therefore, given an application condition ac of a rule p, each occurrence of a subcondition of the form $\exists A$ with A violating one of these EMF constraints, may be replaced by false without altering the meaning. We know that such patterns cannot appear in any EMF instance model. Thus, in the context of EMF, the result is semantically equivalent to the actual guaranteeing rule but may contain fewer subconditions.

Theorem 2 (Correctness of EMF-specific simplifications). *Let c be a graph condition over P and c' be the condition that results from replacing every occurrence of a subcondition $\exists(a : C_1 \hookrightarrow C_2)$ of c by **false** if the graph C_2 contains parallel edges or multiple incoming containment edges to the same node. Then an injective morphism $p : P \hookrightarrow G$ into an EMF-model graph G satisfies c if and only if it satisfies c'. In particular, if c is a graph constraint, any EMF-model graph G satisfies c if and only if it satisfies c'.*

Correctness of this theorem is proven by induction along the nesting structure of the constraint in the cases of parallel edges and multiple containment nodes. The same argument also applies in the case of finite containment cycles. But since containment cycles of arbitrary length cannot be expressed as graph constraints, the correctness of replacing their occurrence by false is intuitive but not amenable to a formal proof by induction.

Example 3 (compare Sect. 2). Theorem 2 would drop the third, sixth, and seventh subcondition from the application condition in Fig. 5 by replacing it with false since it contains a node with more than one container or parallel edges.

5   Tooling

We developed our optimizer as an Eclipse-plugin tool support on top of OCL2AC implementing all of the proposed simplifications except for the elimination of containment cycles. The optimizer consists of two main components: (a) an *analyzer* that detects if a constraint needs to be integrated into a given rule at all (Theorem 1, 1 and 2) and (b) a *simplifier* for eliminating unnecessary subconditions from the guaranteeing application conditions *during the construction process* (Theorems 1, 3 and 2). Given a Henshin rule and a graph constraint, our optimizer automatically renders the rule to preserve the validity of the constraint. Additionally, we implemented simplifications of application conditions by applying well-known equivalence rules like $\exists (C_1, \exists C_2) \equiv \exists C_2$ if $C_1 \subseteq C_2$, $\exists C_1 \vee \exists C_2 \equiv \exists C_1$ if $C_1 \subseteq C_2$, or $\exists C_1 \wedge \exists C_2 \equiv \exists C_2$ if $C_1 \subseteq C_2$ [26]. Applying these, entire graphs may be omitted and even levels of nesting may be collapsed. The tool support can be downloaded from our website[1].

6   Evaluation

In this section, we show the highlights of our evaluation; a comprehensive overview is given in [20] and the artifacts can be downloaded (see footnote 1).

Research Questions (RQs). Our evaluation aims to answer the following RQs regarding the complexity and performance: (RQ 1:) *How complex are the resulting application conditions with and without optimizations? How does this compare to the complexity of the original graph constraints?* To perform validity-preserving steps, there are two basic approaches: We either test for validity after each transformation step and rollback the step if its resulting model is not valid (*a posteriori* check) or the transformation is designed to perform validity-preserving steps only (*a priori* check). We, therefore, ask the following questions: (RQ 2.1:) *How fast is the* a priori *validity check compared to the* a posteriori *check?* (RQ 2.2:) *Does the optimization of application conditions improve the performance significantly?*

General Set-Up. As an application case, we consider the scenario of in-place model transformations that should preserve a basic consistency such that the resulting instances can be opened in a domain-specific model editor throughout. In [16], Kehrer et al. derive consistency-preserving editing rules from a given meta-model. However, they support basic constraints like multiplicities only. More complex OCL constraints are left to future work. In their evaluation, this restriction has the most serious impact on the UML meta-model for Statecharts [25]. Out of 17 original constraints they identified 11 to be enforced in MagicDraw [21]. In total, they used 84 editing rules for Statecharts.

We translated those 11 OCL constraints into graph constraints and then integrated them as application conditions into the 84 rules.

7 valid test models of sizes between 800 to 16 000 elements (nodes and references) are used to conduct our performance experiments. These test models are synthetic containing copies of an initial valid model composing 5 objects of each non-abstract class of the meta-model. All evaluations were performed with a desktop PC, Intel Core i7, 16 GB RAM, Windows 7, Eclipse Neon, Henshin 1.4.

---

[1] https://ocl2ac.github.io/home/.

6.1 Evaluating Complexity

In theory, the size of a computed application condition (the number of graphs) can grow over-exponentially in the worst case compared to the size of the original constraint [26]. In practice, however, the growth is moderate. Mainly due to node typing, many node overlappings are not possible. To find out how far this blow up of application conditions is a problem in practice, we conducted the following experiments considering the number of graphs as well as the number of nesting levels in application conditions. Additionally, we explore how far the complexity can be reduced using our optimization. Table 1 gives an overview of the results.

Integration Without Optimization. Given the 11 OCL constraints of our application case, we translated them to graph constraints containing 2 to 10 graphs (36 in total) and integrated all of these in each of the 84 rules using OCL2AC (i.e., computing the guaranteeing application conditions). The newly added application conditions contain 77.3 graphs on average (with 36 being the best and 191 being the worst case) and 6 nesting levels. Thus, on average the number of graphs more or less doubles which is far better than could be suspected from theory. Nonetheless, the number of graphs is way too high and also the number of levels should be smaller in most cases. Hence, there is a clear need to further optimize the resulting application conditions.

Integration with Optimization. To find out how efficient our optimizations of application conditions are, we conducted the same experiment as above using our developed optimizer. In result, the average number of graphs in the application condition is 10.8 (with 0 being the best and 35 being the worst case), i.e., the complexity is reduced by factor 7 on average using our optimizer. Additionally, the deepest nesting level of 6 was often reduced to at most 2 levels. Theorem 1,1 turns out to be the main reason behind this considerable loss of complexity: Instead of integrating 11 constraints into each rule, on average only 1.7 constraints are integrated into a rule.

Table 1. Number of graphs of application conditions and deepest nesting levels before and after optimization (with emphasis on extreme cases)

Rule	w/o optimization		w optimization		
	#graphs	level	#graphs	level	#integrated constraints
create_Transition	191	6	1	1	1
create_FinalState	44	6	31	6	11
delete_Trigger	37	6	0	0	0
Average (84 rules)	77.3	6	10.8	2.6	1.7

Table 1 shows extreme cases: Considering all 84 rules and the 11 constraints, the best optimization was reached with rule create_Transition where the resulting application condition with 191 graphs was reduced to a condition with just one graph. One of the lowest optimizations came along with rule create_FinalState. Since it is overlapped with all the 11 constraints, the number of the resulting graphs is reduced by factor 1.4 only (using Theorem 1, 3). Rule delete_Trigger started with one of the lowest number of graphs in its application conditions. This condition is eliminated altogether using our optimization.

Across 10 runs, the average time of integrating the 11 graph constraints for statecharts into all 84 rules was 2.3 s. without optimization and 1.03 s. with optimization. In particular, calculation of our simplified application conditions is even faster than computing the guaranteeing ones. In both cases, calculating all needed application conditions for a given rule set is fast enough to be used in practice.

To answer RQ 1, given graph constraints with 2–10 graphs (3.2 on average) and 2–6 nesting levels (2.3 on average), non-optimized application conditions have 36–191 graphs (77.3 on average) and 6 nesting levels, while optimized ones have 0–35 graphs (10.8 on average) and 0–6 nesting levels (2.6 on average). Hence, condition sizes are considerably reduced (by factor 7 on average).

6.2   Evaluating Performance

To answer RQ 2.1 and RQ 2.2, we set up two test scenarios comparing the runtime of *a posteriori* and *a priori* validity checks.

Experiment Set-Up. Each test scenario (TS) consists of 15 test cases, one case for 15 selected rules (out of 84). These 15 rules are representative w.r.t. supported editing actions and rule size, in particular, they cover all kinds of editing actions. Their sizes range between 3 and 7 model elements. The average size of an application condition of the 15 rules is 56.4 graphs with nesting level 6 (without optimization) and 16.8 graphs with nesting level 3.1 (with optimization). A test case of TS 1 consists of first applying an original rule to a test model at a random match and then checking the validity of the resulting model (using (a) the EMF validator [7] configured to employ the OCLinEcore validator [23] to validate OCL constraints and (b) the OCL interpreter [22]). A test case of TS 2 consists of applying an updated rule (with (a) the guaranteeing and (b) the optimized application condition) to a test model at a random match. To eliminate effects stemming from the choice of match, each test case of a test scenario is performed 100 times. A test scenario in TS 1 (a) is performed in one run time session such that caching of information can be used advantageously. A second variant of TS 1 (a) performs each *a posteriori* check in a separate session making caching useless. All the test scenarios have been performed on all the 7 valid test models.

The average run times are measured over altogether 15 000 applications for each scenario. A timeout (TO) takes place if the average run time exceeds 5 min. To evaluate an OCL constraint using the OCL interpreter, the context object has to be given. Focusing on approach differences, the following times were excluded from the evaluation time: The time needed to find the context objects of all

OCL constraints for the OCL interpreter, the loading time of a test model to any validator, and the time needed to roll back to the state of a test model after applying a rule whose resulting model does not satisfy the constraints.

Table 2. Average run time (in seconds) of a single rule application (and validation) over 15 test cases with 100 random matches each using models of varying size

Scenario	(Caching)	Model size						
		800	1 500	3 000	6 000	10 000	13 000	16 000
TS 1(a)	(yes)	0.01	0.01	0.01	0.02	0.04	0.05	0.06
TS 1(a)	(no)	1.66	1.71	1.76	1.79	1.8	1.83	1.85
TS 1(b)	(no)	128.97	185.08	254.17	TO	TO	TO	TO
TS 2(a)	(no)	0.01	0.01	0.04	0.13	0.3	0.5	0.79
TS 2(b)	(no)	0.01	0.01	0.02	0.05	0.12	0.22	0.33

Experiment Results. Table 2 shows the following results: *A posteriori* checking is performed in 3 variants. TS 1 (a) uses the EMF validator with and without caching mechanism since we noted the followings: In the first validation check, the EMF validator took 1.77 to 1.95 s to check a test model of size between 800 to 16 000, whereas in the next validation checks, it took only 5 to 63 ms. Our understanding for this improvement is that the EMF validator saves the model state after the first validity check. Thus, in the next checks at the same run time session, the EMF validator is still able to reach the model in the cache such that only the elements affected by rule application are considered. Without caching, the average run times are less than 2 s; with caching they are even about two magnitudes faster. Using the OCL interpreter (TS 1 (b)) instead leads to run times over 2 min or even timeouts (after 5 min.). *A priori* checking is performed in two variants: In TS 2 (a) rules with non-optimized application conditions are used while the application conditions in TS 2 (b) are optimized. The run times of both variants are below 1 second and hence slightly better than in TS 1 (a) without caching. Using caching, however, TS 1 (a) is even faster. This consideration yields the answer to RQ 2.1. To answer RQ 2.2 we can see that using rules with optimized application conditions is two and a half times faster than without optimization. Almost all of the times our rules were applicable and thus the whole application condition of a rule was completely checked and evaluated. To conclude, we can state that scenarios TS 1 (a) and TS 2 are both fast enough to be usable in practice. However, a rollback step in the *a posteriori* approach (TS 1) may not always be feasible. For example, if the rollback step is defined by applying the inverse rule, this is might not always be applicable if the rule computes attribute values. Furthermore, in the *a posteriori* approach, the rule action is performed first which may cause dangerous situations in several fields such as a railway system, self-driving cars and an e-health system.

Threats to Validity. External validity can be questioned since we consider a limited number of OCL constraints and rules. For our performance experiments, we selected 15 out of 84 editing rules which are representative concerning their kinds (rules for creating, deleting, setting, unsetting, and moving model elements) and sizes. Moreover, we reduced the effect of the rules' matches by executing each rule at 100 matches chosen randomly from each given model. For performance evaluation, we restricted our studies to synthetic models. As we did not spot any performance bottleneck, we are convinced that using realistic models would not yield basically different results.

Concerning the considered OCL constraints it can be noticed that about half of them are simple negative constraints. However, all core features of OCL (logical operators, navigation expressions and collection operators) are covered and at least one rather complex constraint is included. And, more importantly, this kind of constraints seems to be quite typical for the chosen application case. Constraints required by model editors are often negative to forbid input that is not allowed anyway. Therefore, we are confident that the results are representative. Nevertheless, further case examples are interesting to be considered in the future.

7   Related Work

Related works can be distinguished into two groups: (1) other works ensuring transformation rules to be validity-preserving and (2) simplifying (application) conditions and constraints.

Ensuring Transformation Rules to Be Validity-Preserving. In [2,27], Azab, Pennemann et al. introduce ENFORCe, a prototype implementation that can ensure the correctness of graph programs. It integrates graph constraints as left application conditions of rules as well but supports (partially) labeled graphs, not EMF models, and there is no translation from OCL to graph constraints available.

Clarisó et al. present in [5] how to calculate an application condition for a transformation rule and an OCL constraint, directly in OCL. The supported subset of OCL is slightly larger than in OCL2AC because, staying with OCL, they can support operations which are not first-order. The authors provide a correctness proof for the presented translation into application conditions. In addition, there is a partial implementation. Resulting application conditions are not further optimized, neither by ENFORCe nor in the work by Clarisó et al. To the best of our knowledge, our work is the only one which optimizes the resulting application conditions considerably.

Simplifying (Application) Conditions and Constraints. Rules for semantic equivalences in graph constraints and conditions have been reported in several places [26–28] and their application can lead to considerable simplification in the structure of a constraint. There are also approaches and implementations simplifying OCL constraints, especially automatically generated ones [6,12]. Depending on the usage scenario, such simplifications could provide a valuable preprocessing step to our approach.

8   Conclusion

Application scenarios where each graph transformation step has to preserve the validity of models w.r.t. given constraints are needed in practice. As the construction of application conditions in [13] yields validity-guaranteeing ones and assuming that the preservation of graph validity is already sufficient, the resulting application conditions can be considerably optimized. We developed several techniques (in Theorems 1 and 2) to construct optimized validity-preserving application conditions and implemented them on top of OCL2AC. In our evaluation, the usability of OCL2AC was investigated for the first time, with and without optimization. The evaluation results show that OCL2AC can lead to quite large application conditions which can be significantly optimized by factor 7 (on average) using our developed techniques. Accordingly, while the performance results of correct graph transformations are good in general, applying rules with optimized application conditions is shown to be ca. 2.5 times faster than applying non-optimized ones.

In future, we intend to further optimize resulting application conditions by identifying redundant subconditions and by checking negative invariants of modeling languages. Our ultimate goal is to obtain understandable application conditions identifying exactly those portions of the given constraints that are relevant for a given rule. This work is already an essential step into that direction. Moreover, our optimization of conditions could have some interesting applications beyond MDE. We are interested, e.g., in assessing if our ideas can be beneficially integrated into proof systems [27,30].

Acknowledgement. We are grateful to Annegret Habel, Christian Sandmann, and Steffen Vaupel for their helpful comments on a draft version of this paper. This work was partially funded by the German Research Foundation (DFG), projects "Generating Development Environments for Modeling Languages" and "Triple Graph Grammars (TGG) 2.0".

References

1. Arendt, T., Biermann, E., Jurack, S., Krause, C., Taentzer, G.: Henshin: advanced concepts and tools for in-place EMF model transformations. In: Petriu, D.C., Rouquette, N., Haugen, Ø. (eds.) MODELS 2010. LNCS, vol. 6394, pp. 121–135. Springer, Heidelberg (2010). https://doi.org/10.1007/978-3-642-16145-2_9

2. Azab, K., Habel, A., Pennemann, K.H., Zuckschwerdt, C.: ENFORCe: a system for ensuring formal correctness of high-level programs. In: Proceedings of 3rd International Workshop on Graph Based Tools (GraBaTs 2006), vol. 1, pp. 82–93 (2006)

3. Becker, B., Lambers, L., Dyck, J., Birth, S., Giese, H.: Iterative development of consistency-preserving rule-based refactorings. In: Cabot, J., Visser, E. (eds.) ICMT 2011. LNCS, vol. 6707, pp. 123–137. Springer, Heidelberg (2011). https://doi.org/10.1007/978-3-642-21732-6_9

4. Biermann, E., Ermel, C., Taentzer, G.: Formal foundation of consistent EMF model transformations by algebraic graph transformation. Softw. Syst. Model. **11**(2), 227–250 (2012)

5. Clarisó, R., Cabot, J., Guerra, E., de Lara, J.: Backwards reasoning for model transformations: method and applications. J. Syst. Softw. **116**(Suppl. C), 113–132 (2016)
6. Cuadrado, J.S.: Optimising OCL synthesized code. In: Pierantonio, A., Trujillo, S. (eds.) ECMFA 2018. LNCS, vol. 10890, pp. 28–45. Springer, Cham (2018). https://doi.org/10.1007/978-3-319-92997-2_3
7. Eclipse Foundation: Eclipse Modeling Framework (EMF) (2019). http://www.eclipse.org/emf/
8. Egyed, A.: Instant consistency checking for the UML. In: Proceedings of the 28th International Conference on Software Engineering, New York, pp. 381–390 (2006)
9. Ehrig, H., Ehrig, K., Prange, U., Taentzer, G.: Fundamentals of Algebraic Graph Transformation. MTCSAES. Springer, Heidelberg (2006). https://doi.org/10.1007/3-540-31188-2
10. Ehrig, H., Ermel, C., Golas, U., Hermann, F.: Graph and Model Transformation - General Framework and Applications. EATCS. Springer, Berlin (2015). https://doi.org/10.1007/978-3-662-47980-3
11. Giese, H., Glesner, S., Leitner, J., Schäfer, W., Wagner, R.: Towards verified model transformations. In: Proceedings of the 3rd International Workshop on Model Development, Validation and Verification (MoDeV2a), Genova, pp. 78–93 (2006)
12. Giese, M., Larsson, D.: Simplifying transformations of OCL constraints. In: Briand, L., Williams, C. (eds.) MODELS 2005. LNCS, vol. 3713, pp. 309–323. Springer, Heidelberg (2005). https://doi.org/10.1007/11557432_23
13. Habel, A., Pennemann, K.H.: Correctness of high-level transformation systems relative to nested conditions. Math. Struct. Comput. Sci. **19**, 245–296 (2009)
14. Heckel, R., Wagner, A.: Ensuring consistency of conditional graph grammars. Electron. Notes Theor. Comput. Sci. **2**(Suppl. C), 118–126 (1995)
15. Kehrer, T., Kelter, U., Taentzer, G.: Consistency-preserving edit scripts in model versioning. In: 2013 28th IEEE/ACM International Conference on Automated Software Engineering, ASE 2013, pp. 191–201. IEEE, Piscataway (2013)
16. Kehrer, T., Taentzer, G., Rindt, M., Kelter, U.: Automatically deriving the specification of model editing operations from meta-models. In: Van Gorp, P., Engels, G. (eds.) ICMT 2016. LNCS, vol. 9765, pp. 173–188. Springer, Cham (2016). https://doi.org/10.1007/978-3-319-42064-6_12
17. Krause, C., Giese, H.: Probabilistic graph transformation systems. In: Ehrig, H., Engels, G., Kreowski, H.-J., Rozenberg, G. (eds.) ICGT 2012. LNCS, vol. 7562, pp. 311–325. Springer, Heidelberg (2012). https://doi.org/10.1007/978-3-642-33654-6_21
18. Lack, S., Sobociński, P.: Adhesive and quasiadhesive categories. Theor. Inform. Appl. **39**(3), 511–545 (2005)
19. Nassar, N., Kosiol, J., Arendt, T., Taentzer, G.: OCL2AC: automatic translation of OCL constraints to graph constraints and application conditions for transformation rules. In: Lambers, L., Weber, J. (eds.) ICGT 2018. LNCS, vol. 10887, pp. 171–177. Springer, Cham (2018). https://doi.org/10.1007/978-3-319-92991-0_11
20. Nassar, N., Kosiol, J., Arendt, T., Taentzer, G.: Constructing optimized validity-preserving application conditions for graph transformation rules: extended version. Technical report, Philipps-Universität Marburg (2019). https://uni-marburg.de/fb12/arbeitsgruppen/swt/forschung/publikationen/2019/NKAT19-TR.pdf/
21. No Magic: Magic draw. https://www.nomagic.com/products/magicdraw
22. OCL: Eclipse OCL (2019). https://projects.eclipse.org/projects/modeling.mdt.ocl
23. OCLinEcore: Eclipse OCL (2019). https://wiki.eclipse.org/OCL/OCLinEcore

24. OMG: Object Constraint Language (2014). http://www.omg.org/spec/OCL/
25. OMG: OMG Unified Modeling Language. Version 2.5 (2015). http://www.omg.org/spec/UML/2.5/
26. Pennemann, K.H.: Generalized constraints and application conditions for graph transformation systems. Diplomarbeit, Department für Informatik, Universität Oldenburg (2004). https://bit.ly/2T4RV0A
27. Pennemann, K.H.: Development of correct graph transformation systems. Ph.D. thesis, Carl von Ossietzky-Universität Oldenburg (2009)
28. Radke, H., Arendt, T., Becker, J.S., Habel, A., Taentzer, G.: Translating essential OCL invariants to nested graph constraints for generating instances of metamodels. Sci. Comput. Program. **152**, 38–62 (2018)
29. Rensink, A.: Representing first-order logic using graphs. In: Ehrig, H., Engels, G., Parisi-Presicce, F., Rozenberg, G. (eds.) ICGT 2004. LNCS, vol. 3256, pp. 319–335. Springer, Heidelberg (2004). https://doi.org/10.1007/978-3-540-30203-2_23
30. Schneider, S., Lambers, L., Orejas, F.: Automated reasoning for attributed graph properties. Int. J. Softw. Tools Technol. Transf. **20**(6), 705–737 (2018)
31. Sendall, S., Kozaczynski, W.: Model transformation: the heart and soul of model-driven software development. IEEE Softw. **20**(5), 42–45 (2003)
32. Steimann, F., Frenkel, M., Voelter, M.: Robust projectional editing. In: Proceedings of the 10th ACM SIGPLAN International Conference on Software Language Engineering. SLE 2017, pp. 79–90. ACM, New York (2017)

From Pattern Invocation Networks
to Rule Preconditions

Nils Weidmann[1]([⊠]), Anthony Anjorin[1], Florian Stolte[2], and Florian Kraus[1]

[1] Paderborn University, Paderborn, Germany
{nils.weidmann,anthony.anjorin,florian.kraus}@upb.de
[2] itemis AG, Paderborn, Germany
fstolte@itemis.com

Abstract. Incremental (graph) pattern matchers provide a suitable, high-level platform for implementing Graph Transformation (GT) engines. All incremental pattern matchers we are aware of use a similar notion of Pattern Invocation Networks (PINs) as a specification language. Leveraging an incremental pattern matcher for GT thus requires a semantics-preserving transformation from GT rules to PINs. Although graph queries have been formally related to generalised discrimination networks (a generalisation of PINs) in the literature, practical GT engines typically support only a much more restrictive form of "flat", i.e., non-nested graph queries. We are not aware of any formalisation that relates PINs to non-nested graph queries in a way that supports verifying semantics preservation for GT-to-PIN transformations and PIN-to-PIN optimisations in a fully automated manner. In this paper, we therefore propose a formal semantics for a specific class of "flat-equivalent" PINs by providing a flattening transformation to non-nested graph queries.

Keywords: Graph constraints · Pattern Invocation Networks · Incremental pattern matching

1 Introduction and Motivation

Graph pattern matchers (PMs) are essential core components for Model-Driven Engineering (MDE) tools as they enable an abstraction from details of model traversal. *Incremental* PMs provide the additional advantage of efficiently keeping track of all available (partial) matches for a given set of patterns. As matches are not calculated on demand but instead always maintained in memory, matches can be updated efficiently instead of being recalculated from scratch when the model changes, e.g., due to the application of a rule or due to changes made by a user. This is particularly effective for small changes in large models, or when most of the matches are eventually required for the transformation [4,15].

When building a Graph Transformation (GT) tool on top of an incremental PM, the preconditions of large GT rules can be decomposed into smaller subpatterns connected in a network-like structure called a Pattern Invocation

© Springer Nature Switzerland AG 2019
E. Guerra and F. Orejas (Eds.): ICGT 2019, LNCS 11629, pp. 195–211, 2019.
https://doi.org/10.1007/978-3-030-23611-3_12

Network (PIN). If this decomposition is *suitable*, runtime and memory consumption can be reduced by reusing and sharing common partial matches across the network [18]. The challenge here is determining what constitutes a suitable PIN in relation to the size and structure of the involved (meta)models, patterns, and the specific incremental PM. While the task of finding an optimal PIN can be fully delegated to end users, we believe that a (perhaps configurable) automation of this process is desirable in most cases. To accomplish such an automation, a GT tool developer must thus program and evaluate *various* decomposition strategies into PIN structures for the *same* pattern. This, however, raises a new question related to the correctness of a decomposition strategy: *Is a given PIN "equivalent" to the original "flat" pattern before the decomposition?*

To support GT tool developers by automating the verification of PIN structures, we propose an algorithm for transforming a PIN into a unique flattened form for a given root node, which can then be easily identified with the original pattern. Our algorithm can be used to support the verification of decomposition strategies and PIN-to-PIN optimisations.

The rest of this paper is structured as follows: An intuition for basic concepts such as GT rules, rule preconditions and PINs is given in Sect. 2. To establish a formal underpinning for working with PINs, we formalise the syntax of PINs based on graphs and graph morphisms in Sect. 3. In Sect. 4, we provide a PIN semantics via an algorithm that flattens a PIN with a distinguished root to a unique form (Algorithm 1). We then map this flattened form to a rule precondition (graph constraint) in Algorithm 2, establishing *graph constraints* from the framework of algebraic graph transformation as a semantic domain for PINs. The runtime performance of Algorithm 1 is evaluated in Sect. 5 with a case study. In Sect. 6 we provide an overview of related work, before concluding and proposing future work in Sect. 7.

2   Running Example and Basic Concepts

Our example is inspired by the *FamiliesToPersons* case, which has been used in many variants to benchmark model transformation tools (cf. e.g. [1]). Figure 1 depicts the *Persons* metamodel. A person register contains multiple persons (male or female), while a person can be responsible for another person, in the sense of having a power of attorney for them.

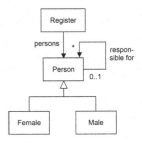

Fig. 1. Metamodel

In order to manipulate graphs, *rules* are used for adding and deleting nodes and edges. Figures 2 and 3 depict two rules used to manipulate persons models. Black elements represent the required context for applying a rule, grey elements represent Positive Application Conditions (PACs) that extend this context and must be present for the rule to be applicable. Blue elements (with a "!" mark-up) represent Negative Application Conditions (NACs), i.e., a rule cannot be applied if these elements are present. Green elements (with a "++" mark-up) are to be created, and red elements (with a "--" mark-up) deleted. To simplify the diagrams, *responsible for* and *persons* are abbreviated with *rF* and *per*, respectively.

The rule *AddResponsibility* adds an *rF* edge between two persons *p1* and *p2* of the same person register, but only if this edge does not already exist (first NAC) and if no other person *p3* is already responsible for *p2* (second NAC). *DeletePerson* deletes a person *p2* together with a *per* edge and an *rF* edge, if *p2* is not responsible for any other person *p3* (NAC).

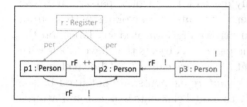

Fig. 2. *Rule: AddResponsibility* (Color figure online)

Fig. 3. *Rule: DeletePerson* (Color figure online)

In this paper, we are primarily interested in *pattern matching*, i.e., determining a valid assignment (a *match*) of all elements required by a rule to model elements in a host graph, but not actually in applying the rule. For this reason, we shall focus in the following on the *precondition* of a rule, consisting of all black and red elements, as well as all PACs and NACs.

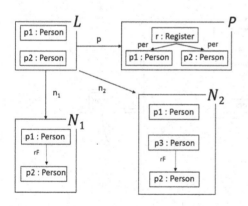

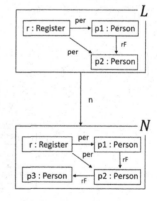

Fig. 4. Precondition for *AddResponsibility*

Fig. 5. Precond. for *DeletePerson*

Figures 4 and 5 depict the preconditions for *AddResponsibility* and *Delete-Person* in a formal notation (all definitions are provided in Sect. 3) with top-level objects as typed graphs, and top-level arrows as graph morphisms. The exact mappings for all graph morphisms are indicated by the node labels in the diagrams (e.g., person $p1$ in L is mapped by arrow p to person $p1$ in P). L denotes the left-hand side of each rule (all black and red elements), P the PAC for *AddResponsibility*, N the NAC for *DeletePerson*, and N_1, N_2 the NACs for *AddResponsibility*.

Figure 6 depicts a PIN for maintaining all matches for the rules *AddResponsibility* and *DeletePerson*. The top-level objects in the diagram are referred to as *patterns*, connected by either *positive invocations* (black arrows), or *negative invocations* (dashed, blue arrows). When a pattern is positively invoked, this means that the invoking pattern can only match if the invoked pattern matches; A match of a negatively invoked pattern prevents the invoking pattern from matching. The mappings for invocation arrows are indicated via labels on the arrows of the form `from → to`. Inside the patterns, objects that have labels with a bar (such as $\overline{p3}$, $\overline{r}$) and all their incident arrows are called *local*. Local elements, as opposed to all other elements called *signature* elements, are not fixed by the invoking pattern and can be bound freely to model elements in a match. Patterns without incoming edges are called *root patterns*. Given a host graph, a PM maintains all matches for every pattern in the PIN whose positively invoked patterns also match, and whose negatively invoked patterns do not match. Matches for root patterns are reported by the PM as *complete* matches. All other matches are *partial* and are only computed and maintained internally.

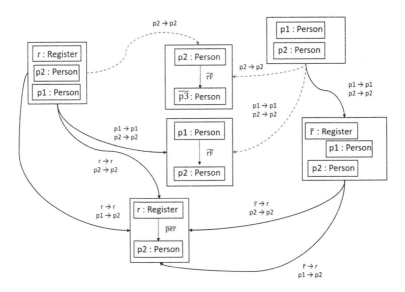

Fig. 6. Combined pattern invocation network for both rules (Color figure online)

The PIN depicted in Fig. 6 has two root patterns corresponding to the preconditions for our two rules. As some patterns are *shared*, i.e., invoked by multiple patterns, an incremental PM might be able to reuse partial matches and reduce the overall pattern matching effort required to maintain all matches. Whether this PIN is, however, actually advantageous in practice depends not only on how the specific incremental PM works internally, but also on the patterns and models. As an example, although pattern sharing can be increased by reducing the size of individual patterns, this often increases the number of partial matches that are collected but later discarded when combining partial matches to form complete matches, with a negative effect on runtime. Figure 7 depicts separate PINs for both rules that avoid pattern sharing completely. These two PINs are equivalent to the single PIN depicted in Fig. 6 in the sense that exactly the same set of complete matches are maintained for the two rules.

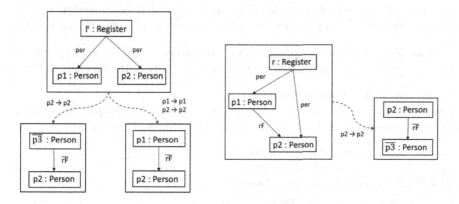

Fig. 7. Flat PIN for *AddResponsibility* (left) and *DeletePerson* (right)

3   Semantics of Pattern Invocation Networks

As we aim to formalise a pattern as a *graph morphism* embedding its signature in its body, we introduce the basic concepts of graphs and graph morphisms:

Definition 1 ((Typed) Graph and (Typed) Graph Morphism).
A graph $G = (V, E, src, trg)$ consists of a finite set V of nodes, a finite set E of edges, and two functions $src, trg : E \to V$ that assign each edge a source and target node, respectively. The set $elts(G) = V \mathbin{\dot\cup} E^1$ denotes the union of nodes and edges. Given graphs $G = (V, E, src, trg)$, $G' = (V', E', src', trg')$, a graph morphism $f : G \to G'$ consists of two total functions $f_V : V \to V'$ and $f_E : E \to E'$ such that $src \mathbin{;} f_V = f_E \mathbin{;} src'$ and $trg \mathbin{;} f_V = f_E \mathbin{;} trg'$. The $\mathbin{;}$ (then) operator used here denotes the composition of functions: $(f \mathbin{;} g)(x) := g(f(x))$.
A graph monomorphism is a graph morphism $f = (f_V, f_E)$ with injective functions f_V and f_E, denoted by $f : G \hookrightarrow G'$. A type graph is a distinguished graph $TG = (V_{TG}, E_{TG}, src_{TG}, trg_{TG})$. A typed graph is a pair $\hat{G} = (G, type)$ of a

---

[1] $\dot\cup$ denotes the disjoint union of sets, $\emptyset$ the empty set.

graph G together with a graph morphism type : G → TG. Given typed graphs
$\hat{G} = (G, type)$ *and* $\hat{G}' = (G', type')$, *a typed graph morphism* $f : \hat{G} → \hat{G}'$ *is a*
graph morphism $f : G → G'$ *such that* $f ; type' = type$.

To simplify the presentation, everything in the rest of the paper is *typed* unless
we explicitly state otherwise, i.e., we write graph but mean *typed* graph.

In analogy to a method with arguments and local variables, a *pattern* in
the context of PINs consists of two disjoint sets of signature elements and local
elements, which together make up the body of the pattern:

Definition 2 (Pattern).
A pattern $p : S \hookrightarrow B$ *is a graph monomorphism with source graph S (called*
its signature*), and target graph B (called its* body*). The nodes and edges of S*
are called signature elements*, the nodes and edges of B* body elements*. Body*
elements that are not in $p(S)$ *are called* local elements*, i.e.,* $elts(B) \setminus elts(p(S))$.

In analogy to method invocation, one pattern can invoke another by embedding the signature of the invoked pattern into the body of the invoking pattern:

Definition 3 (Pattern Invocation).
A pattern invocation i *from a pattern* $p : S \hookrightarrow B$ *to a pattern* $p' = S' \hookrightarrow B'$ *is*
a graph monomorphism $e : S' \hookrightarrow B$, *which we denote by* $i : p → p'$.

Example 1 The figure to the right
depicts the PIN for *DeletePerson*
from Fig. 7 in formal notation. The
root pattern $p : S \hookrightarrow B$ has no local
elements so its signature and body
are identical. The pattern $p' : S' \hookrightarrow$
B' only has one of the persons in
its signature as the other person is a
local node of the pattern. The pattern
invocation $i : p → p'$ is formalised
with the graph monomorphism $e :$
$S' \hookrightarrow B$. Note how the "arrows" i and
e go in different directions as the signature of the invoked pattern must be
embedded in the body of the invoking
pattern.

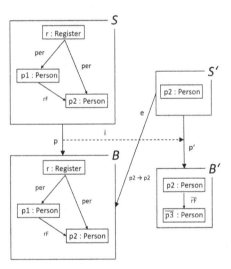

Definition 4 (Pattern Invocation Network (PIN)).
A PIN is a directed graph (Pt, I, src, trg) *with a set Pt of patterns as nodes and*
a set $I = I^+ \,\dot{\cup}\, I^-$ *of pattern invocations as edges, where* $\forall i \in I.\ i : p → p' \Leftrightarrow$
$src(i) = p \wedge trg(i) = p'$. *We refer to* I^+ *as the positive pattern invocations,* I^-
as the negative pattern invocations, $\{rt \in Pt \mid \nexists i \in I.\ trg(i) = rt\}$ *as the roots,*
and $\{lf \in Pt \mid \nexists i \in I.\ src(i) = lf\}$ *as the leaves of the PIN. We also write* (Pt, I)
for (Pt, I, src, trg) *if the src and trg functions are not directly relevant.*
A PIN is flat iff it has a single root rt, $\forall i \in I.\ src(i) = rt$, *and* $I^+ = \emptyset$.

To flatten a PIN, we shall systematically *merge* invoked patterns with their invoking patterns starting with the leaves of the network, i.e., patterns that do not invoke any other patterns. This process of merging pattern bodies, whilst taking signature elements into account, can be formalised as building a *pushout*.

Definition 5 (Pushout).
The pushout *of graph monomorphisms* $e : S' \hookrightarrow B$ *and* $p' : S' \hookrightarrow B'$ *is defined by* (B^*, p^*, e^*) *where:* $p^* : B \to B^*$ *and* $e^* : B' \to B^*$ *are graph morphisms such that the "pushout square" commutes, i.e.,* $e; p^* = p'; e^*$, *and* $\forall (B^\#, p^\# : B \to B^\#, e^\# : B' \to B^\#)$ *with* $e; p^\# = p'; e^\#$, *there exists a unique* $x : B^* \to B^\#$ *such that* $(p^*; x = p^\#) \wedge (e^*; x = e^\#)$. *This "binary" pushout construction can be generalised to a* multi-pushout *[12] of a finite set of morphisms.*

Not all PINs according to Definition 4 can be flattened, i.e., are in this sense equivalent to a flat PIN according to Definition 4. With the following definition, we thus characterise a subset of all PINs, namely *flat-equivalent PINs*, which fulfil a set of conditions to ensure that they can be flattened by our algorithm.

Definition 6 (Flat-Equivalent PIN).
A PIN is flat-equivalent *if it satisfies the following three conditions:*
(C1) It is a directed acyclic graph (DAG), (C2) there exists no directed path in the PIN containing more than one negative pattern invocation, and (C3) for every pattern invocation $i : p \to p'$ *via* $e : S' \hookrightarrow B$ *in the PIN, there exists no local node* $\bar{n}$ *in* B' *that is of the same type of a node* $n \in B \setminus e(S')$.

The first condition (C1) ensures termination by ruling out invocation cycles in PINs. Although cyclic PINs can be assigned a meaningful semantics (cf. recursive pattern matching [19]), we leave this extension to future work. The second condition (C2) forbids nested negation, as we choose our semantics domain to cover only NACs and not general *nested* graph conditions [10]. The third condition (C3) ensures that the PIN can be represented as a single pattern that can be matched *injectively* (cf. Definition 7). If non-injective matching is desired then this condition can be omitted.

The final step in our flattening transformation is to interpret flat PINs as *preconditions* for graph transformation rules (cf. Figures 4 and 5):

Definition 7 (Rule Precondition).
A rule precondition $pre = (L, p : L \to P, \mathcal{N})$ *consists of a graph* L, *a monomorphism* $p : L \hookrightarrow P$ *(a PAC), and a set* $\mathcal{N} = \{n_i : L \hookrightarrow N_i \mid i \in I\}$ *of monomorphisms (NACs) for an index set* I. *For a given graph* G, *an arrow* $m : L \hookrightarrow G$ *satisfies pre, denoted by* $m \models pre$, *iff* $[\exists m_p : P \hookrightarrow G.\ m = p ; m_p] \wedge [\forall i \in I. \nexists m_{n_i} : N_i \hookrightarrow G.\ m = n_i ; m_{n_i}]$, *where* $m, m_p, (m_{n_i})_{i \in I}$ *are monomorphisms.*

4   The Flattening Algorithm

We provide a formal semantics for PINs via two algorithms FLATTENNETWORK and CREATERULEPRECONDITION. The former converts a flat-equivalent PIN into a flat PIN, while the latter interprets the flat PIN as a rule precondition.

FLATTENNETWORK (Algorithm 1), takes as input a flat-equivalent PIN and one of its roots, and produces a flat PIN. It does this in three loops: the first loop from Lines 6–10 recursively flattens all invocations of the chosen root, passing the invoked child pattern ($child_i$) as the new root for the algorithm.

As depicted in Fig. 8, the second loop from Lines 12–20 merges each positively invoked and now flattened child with the root pattern. The inner loop from Lines 16–19 transfers all negative invocations of the child to the pattern that results from the merge operation. On Lines 25–26, a multi-pushout (depicted in Fig. 9) is used to combine all intermediate merge results into a single root pattern $root^*$. In the final loop from Lines 29–33, all negative invocations are pulled up to this new root, resulting in the flat PIN returned on Line 34.

Algorithm 1. FLATTENNETWORK(PIN, $root$)

1: **input:** (1) A flat-equivalent PIN $= (Pt, I)$, and
2:     (2) A chosen root of PIN, $root : S \hookrightarrow B, root \in Pt$
3: **output:** (1) A flat PIN$^* := (Pt^*, I^*)$, and
4:     (2) The unique root of PIN $^*, root^* \in Pt^*$
5: Let $I' := \emptyset, Pt' := \{root\}$
6: **for all** invocations $i \in I, i : root \to child_i$ **do**
7:     $[(Pt_i, I_i), child_i^*] := $ FLATTENNETWORK(PIN, $child_i$)
8:     $Pt' := Pt' \cup Pt_i$
9:     $I' := I' \cup I_i \cup \{i^* : root \to child_i^*\}$
10: **end for**
11: Let PIN$' := (Pt', I' = I^+ \dot{\cup} I^-), R^\# = \emptyset, I^\# = \emptyset$
12: **for all** $i \in I^+, i : root \to child_i$ via $e_i : S_i \hookrightarrow B$ **do**
13:     $(B_i^\#, child_i^\#, e_i^\#) := $ pushout$(S_i, e_i, child_i)$ ▷ See Fig. 8
14:     $root_i^\# := root \,; child_i^\#$
15:     $R^\# := R^\# \cup \{root_i^\#\}$
16:     **for all** $j \in I^-, j : child_i \to neg_{i,j}$ via $n_{i,j} : S_{i,j} \hookrightarrow B_i$ **do**
17:         $n_{i,j}^\# := n_{i,j} \,; e_i^\#$ ▷ See Fig. 8
18:         $I^\# := I^\# \cup \{i^\# : root_i^\# \to neg_{i,j}$ via $n_{i,j}^\# : S_{i,j} \hookrightarrow B_i^\#\}$
19:     **end for**
20: **end for**
21: Let $R^\# = \{root_1^\#, \ldots, root_m^\#\}$
22: **if** $|R^\#| = 0$ **then** $root^* := root$
23: **else if** $|R^\#| = 1$ **then** $root^* := root_1^\#$
24: **else**
25:     $(B^*, root_1^*, \ldots, root_m^*) := $ pushout$(S, root_1^\#, \ldots, root_m^\#)$ ▷ See Fig. 9
26:     $root^* := root_1^\# \,; root_1^*$
27: **end if**
28: Let $Pt^* = \{root^*\}, I^* = \{i \in I^-, i : root \to child_i$ via $e_i : S_i \hookrightarrow B\}$
29: **for all** $i^\# \in I^\# : root_i^\# \to neg_{i,j}$ via $n_{i,j}^\# : S_{i,j} \hookrightarrow B_i^\#$ **do**
30:     $n_{i,j}^* := n_{i,j}^\# \,; root_i^*$ ▷ See Fig. 9
31:     $I^* := I^* \cup \{i^*, i^* : root^* \to neg_{i,j}$ via $n_{i,j}^* : S_{i,j} \hookrightarrow B^*\}$
32:     $Pt^* := Pt^* \cup \{neg_{i,j}\}$
33: **end for**
34: **return** $[(Pt^*, I^*), root^*]$

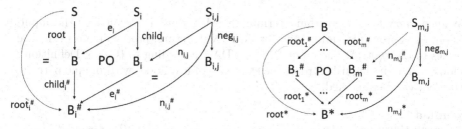

Fig. 8. Merging positive invocations   **Fig. 9.** Creation of new root pattern

Example 2.   Figure 10 depicts a PIN for the rule *AddResponsibility* (cf. Fig. 2). The root pattern with body B, positively invokes two patterns with bodies B_1 and B_2, and negatively invokes (indicated by dashed blue lines) two patterns with bodies B_3 and B_4. Applying FLATTENNETWORK to this PIN and its single root pattern results in all child patterns being converted into flat PINs. For this simple example, however, this is already the case in the initial PIN.

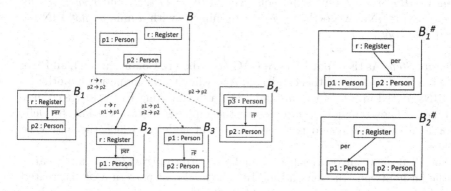

Fig. 10. Initial PIN (Color figure online)   **Fig. 11.** Resulting intermediate bodies (cf. Fig. 8)

As a result of the pushout construction on Line 13, two intermediate bodies $B_1^{\#}$ and $B_2^{\#}$ are constructed (Fig. 11), with a *per* edge inserted between the register r and the persons $p2$ and $p1$, respectively. In a last step, a multi-pushout (Line 25) is constructed from the two intermediate bodies, and connected to the negatively invoked patterns (final loop). Figure 12 depicts the resulting flat PIN.

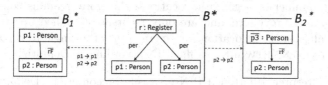

Fig. 12. Flattened network

We now show that Algorithm 1 terminates with a unique result for valid input. Lemma 1 is proven by induction over the PIN structure, proceeding from the leaf nodes to the root. As the input PIN is a DAG according to (C1) of Definition 6 along with a distinguished root, the existence of leaf nodes with unique paths to the root can be assumed.

Lemma 1.
FLATTENNETWORK *(Algorithm 1) produces a unique, flat PIN for valid input.*

Proof (Sketch).

Base Case: Let PIN $= (\{root\}, \emptyset)$ be a PIN which only consists of a single pattern as root and leaf. As leaf patterns do not have invocations, the loop from Lines 12 to 20 is skipped, $|R^{\#}| = |\emptyset| = 0$, and thus $root^*$ is set to $root$ on Line 22. As $I^{\#}$ is empty, the loop from Lines 29 to 33 is skipped, and FLATTENNETWORK returns the input PIN $=$ PIN*, which is already flat according to Definition 4.

Induction Hypothesis: Given flat-equivalent PIN $= (Pt, I)$, and $root \in Pt$ as input, FLATTENNETWORK(PIN, $child_i$) terminates with a unique, flat PIN for all $i \in I, i : root \to child_i$.

Inductive Step: As the input PIN is a DAG according to Condition (C1), all PINs rooted in $(child_i)_{i \in I}$ fulfil this condition. According to the induction hypothesis, the PINs rooted in $(child_i)_{i \in I}$ are flat PINs after invoking FLATTENNETWORK on Line 7. The recursion *terminates* as the input is finite according to Definition 1, and every directed path ends in a leaf node for which the base case applies.

For all positive invocations, intermediate patterns $(root_i^{\#})_{i \in I^+}$ are created via the pushout construction on Line 13. The final root pattern $root^*$ is created from these intermediate root patterns by the multi-pushout construction on Line 25. In both cases, the uniqueness of the pushout object [7] guarantees that the result is unique, i.e., is independent on the order in which patterns are merged.

All negative invocations to the final root pattern $root^*$ are created by concatenating the arrows $(n_{i,j}^{\#})_{i \in I^+, j \in I^-}$ to the intermediate root patterns $(root_i^{\#})_{i \in I^+}$, and the arrows $(root_i^*)_{i \in I^+}$ induced by the multi-pushout construction (Line 30). As Condition (C2) forbids nested negative invocations, i.e., there exists at most one negative invocation on each path from root to leaf, negatively invoked patterns cannot invoke further patterns at this point. The set I^* contains one invocation i^* for each $i^{\#} \in I^{\#}$, which itself contains one invocation for each $j \in I^-$. The number of negative invocations, therefore, remains constant during the flattening procedure, and are only recursively "pulled up" to the root pattern. As all positive invocations are merged into the single root pattern, the output of FLATTENNETWORK is a flat PIN according to Definition 7.   □

We now interpret a flattened PIN as a rule precondition (Definition 7), via the following algorithm CREATERULEPRECONDITION (Algorithm 2):

Algorithm 2. CREATERULEPRECONDITION(PIN)

1: **input:** A flat $PIN = (Pt, I = I^-)$ with $root \in Pt, root : S \hookrightarrow B$
2: **output:** A rule precondition $pre := (L, p : L \rightarrow P, \mathcal{N})$
3: $P := B, L := S, \mathcal{N} := \emptyset$
4: **for all** $i \in I : root \rightarrow neg_i$ via $e_i : S_i \hookrightarrow B$ **do**
5:     $(N_i, n_i, e_i') = $ pushout(S_i, e_i, neg_i)
6:     $\mathcal{N} := \mathcal{N} \cup \{n_i\}$
7: **end for**
8: **for all** $elem \in B \setminus S$ **do**
9:     **if** $\exists i : root \rightarrow neg_i \in I$ via $e_i : S_i \hookrightarrow B$ such that $elem \in e_i(S_i)$ **then**
10:         $L := L \cup \{elem\}$
11:     **end if**
12: **end for**
13: $p(x) := \begin{cases} root(x) : & x \in S \\ id_B(x) : & x \in B \setminus S \end{cases}$
14: **return** $(L, p, \mathcal{N})$

Each negative invocation of the PIN is interpreted as a NAC in the rule precondition by merging the bodies of the invoked patterns into the root (Lines 4–7). To determine L as the minimal context of the precondition, we take all signature elements S, but have to include additional elements from $B \setminus S$ that are passed to a negatively invoked pattern (Lines 8–12).

Example 3.   To demonstrate CREATERULEPRECONDITION, we apply it in the following to the result of Example 2. Via the pushout construction on Line 5, the two NACs N_1 and N_2 are created as depicted in Fig. 13. After creating these NACs, the context L for the precondition is determined (Lines 8–12). In this simple case, L is exactly the signature of the root pattern and all local elements can be moved to the PAC P of the precondition. The resulting precondition is exactly what was discussed in the motivation (cf. Fig. 4).

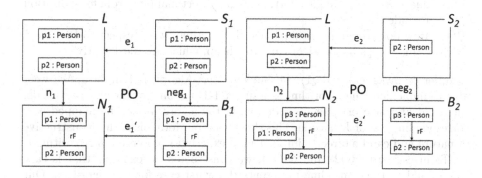

Fig. 13. NACs generated by pushout construction

We show with the following Lemma 2 that CREATERULEPRECONDITION produces a unique rule precondition for any flat pattern:

Lemma 2.
CREATERULEPRECONDITION *(Algorithm 2) terminates with a unique rule precondition for any flat PIN.*

Proof (Sketch). The pushout construction used to create a NAC from a negative invocation on Line 5 guarantees that the result $n_i : L \to N_i$ is a unique monomorphism for every $i \in I^-$ [7]. In the loop from Lines 8–12, the context L is formed by extending the signature by any local elements that are used in a negative invocation. L is well-defined, i.e., a graph, because $e_i : S_i \hookrightarrow B$ is a graph morphism and is thus structure preserving. It is thus impossible to add edges to L without adding their incident nodes. As I and B are finite, the algorithm terminates with a unique rule precondition according to Definition 7. □

With Algorithms 1 and 2, we can now provide a semantics for flat-equivalent PINs:

Definition 8 (Semantics of Flat-Equivalent PINs).
A flat-equivalent PIN = (Pt, I) with chosen root pattern $rt \in Pt$, is semantically equivalent to the rule precondition $(L, p, \mathcal{N})$ = CREATERULEPRECONDITION(FLATTENNETWORK(PIN, rt)).

5   Evaluation

To be of practical use, our proposed algorithm must be (i) implementable with reasonable effort in a mainstream language, and (ii) must produce results in acceptable time for a GT tool developer. These challenges lead to the following research questions, which we shall investigate in a subsequent evaluation:

(RQ1) Can our algorithm be implemented in a mainstream programming language with reasonable effort, i.e., via a direct mapping from formalisation to code?

(RQ2) How does the implementation scale (runtime) with respect to PIN size?

(RQ3) Which steps in the algorithm are cheap, which are most costly?

In order to investigate (RQ1), we implemented Algorithm 1 in Java. We were able to implement the algorithm as a direct 1-1 mapping of the theory to code, such that almost every pseudo code line in the algorithm can be mapped to a corresponding line of Java code. This shows that our formalisation is constructive in nature and can be directly implemented exactly as presented in Algorithm 1.

To investigate (RQ2) and (RQ3), we constructed an example that is easy to test and at the same time represents the worst case for the algorithm. Our experiments revealed that "deep" pattern networks with a long chain of positive invocations and negative invocations on the lowest level represent the worst case for the algorithm, whereas for "broad" networks (smaller maximum depth of sub trees, negative invocations close to the root, such they need not be pulled up often), a better runtime performance was shown for the same number of patterns. In comparison, the actual size (number of nodes and edges) of each pattern is less important. Our chosen example is thus a series of pattern networks for linked lists of increasing length. The recursion hierarchy thus grows linearly with the length of the list, and all negative invocations have to be pulled up from each level to the root. We measured the time required for flattening PINs of increasing size, partitioned into four main steps of the algorithm.

Figure 14 depicts the minimalistic metamodel of a directed graph, which is sufficient for our example. Figure 15 depicts a PIN for a linked list of length 5. All nodes are of type *Node*. Each pattern consists of two nodes connected by a *next* edge. The positive invocations link the source nodes of the invoking patterns to the target nodes of the invoked patterns. The negative

Fig. 14.   Metamodel for a directed graph

invocations prevent each node from having other outgoing edges, such that the pattern matches for a single linked list and not a tree. The flattened network for this example is depicted in Fig. 16. The positively invoked patterns were merged into one list, whereas all negative invocations were recursively pulled up to the root node.

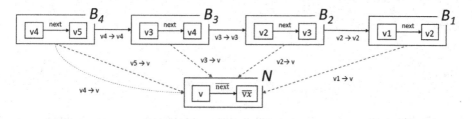

Fig. 15. PIN for a linked list of 5 nodes

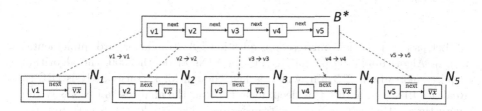

Fig. 16. Flattened PIN for size 5

Experiment Setup: Time measurements were conducted for flattening PINs matching linked lists of length 100 to 5000 in steps of 100. The algorithm was implemented as a Java 1.8 application, executed on a Windows machine using an

Intel Core i7 processor with 16 GB main memory, of which 4 GB were reserved for the JVM. The stack size was set to 16 MB. To reduce the effect of outliers, the median values of three test runs were taken. Details of the runtime measurements[2] and the source code[3] are available online.

Results: The measured runtime values are shown in Fig. 17. Each curve (1)–(4) represents the time consumption of a different part of the code, while the bold, blue line depicts the total runtime (cf. caption for details). Most time is consumed by the second and third steps, whereas the runtime of the first and fourth step is almost negligible. This indicates that the multi-pushout construction is the most expensive task, whereas concatenation of arrows and set operations do not have a significant impact on performance. While the overall runtime is just a few seconds for lists with hundreds of nodes, it exhibits roughly cubic growth and attains 40 min for lists with 5000 nodes.

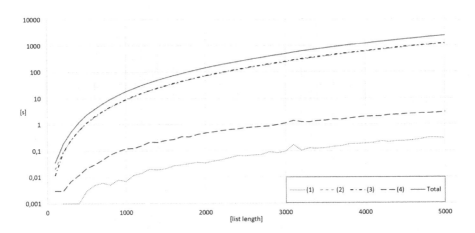

Fig. 17. Runtime: (1) Collect flattened patterns and invocations after recursive call (ll. 6–10), (2) Merge positively invoked patterns and pull up negative invocations (ll. 12–20), (3) Merge intermediates (ll. 22–27), (4) Construct flat PIN (ll. 29–33) (Color figure online)

Let us briefly revisit our research questions: A straightforward implementation of Algorithm 1 is possible (RQ1). For PINs of depth 2000, the algorithm requires only about 2–3 min, which is fast enough for a static analysis (RQ2). Realistic PINs are hardly this deep, meaning that the analysis can be easily run every time a PIN is generated or transformed. The bottleneck is certainly the pushout construction (RQ3); this would thus be a good candidate to be highly optimised and provided in a generic library for constructing GT-based static analyses.

---

[2] bit.ly/2VUWwjM.
[3] github.com/eMoflon/pin-analysis-icgt2019.

6   Related Work

Incremental (graph) pattern matching is an important enabling technology for simplifying the development of various MDE tools: Democles [18] has been used for model synchronisation [15], Drools[4] has been used for model transformation [2,8], implementing a collaborative, event-based modelling framework [20], and providing recommendations via auto-completion [14], Viatra [17] has been used in numerous projects requiring reactive, and event-based programming. All such approaches leveraging incremental pattern matchers can profit from our formalisation as it provides a mapping to algebraic graph transformations for which numerous analysis techniques exist [7]. Tools that support *nested* NACs (e.g., Viatra), however, would require a corresponding extension of our approach.

Furthermore, for tools such as Viatra [17] that allow the end user to directly influence the structure of the PIN, our flattening algorithm can still be used to give feedback by showing the semantically equivalent flat pattern. For tools that derive *multiple* patterns from a single specification [15], however, the process must be fully automated and our algorithm can help to verify the correctness of (configurable) decomposition strategies provided by the tool developers.

Similar to our work, Beyhl et al. [6] show that *Generalised Discrimination Networks* (GDNs) [11], a generalisation of PINs, have the same expressive power as nested graph conditions [10]. As both GDNs and PINs are nested structures, however, the mapping provided by Beyhl et al. [6] cannot be applied to transform PINs to our flat rule preconditions. Providing a mapping to simpler (flat) graph conditions as we do is, however, important in practice as many graph transformation tools do not support the full power of nested graph conditions [16]. This is due to the high price for this expressiveness: guaranteeing scalability becomes difficult, and the analysis (satisfiability, tautology and equivalence) of nested graph conditions is undecidable in general [10].

Finally, while there has been a substantial amount of work on establishing algorithms and extensions for incremental graph pattern matching [3,5,13,18, 19], none of these works provide a formalisation that can be used to bridge graph transformations and incremental graph pattern matchers.

7   Conclusion and Future Work

In this paper, we proposed a formal semantics for flat-equivalent PINs by providing a flattening transformation (FLATTENNETWORK) and a subsequent conversion (CREATERULEPRECONDITION) to a rule precondition. For both the flattening and conversion algorithms, we show termination and uniqueness for valid input. We have implemented our approach via a straightforward mapping of the algorithm to a program in a mainstream programming language. The results indicate that the only expensive part of the implementation is the pushout construction for merging patterns. The overall runtime of a few seconds for up to 1000 recursive calls should be acceptable for most practical applications.

---

[4] www.drools.org.

While our presentation uses typed, attributed graphs with node inheritance, our formalisation is actually generic in the sense that our concepts for *patterns*, *pattern invocations*, and most steps in our algorithms only require a category for which the pushout construction is defined. As we do not fix the objects and arrows in this category, our formalisation could be easily transferred to and used for finite diagrams of graphs, hypergraphs, and many other structures.

As future work, our constructive formalisation could be extended to cover nested negative invocations *of a fixed maximum length*, e.g., consisting of a premise and conclusion (nesting level of one). This could lead to a better compromise between expressiveness, analysability, and scalability. It would also be interesting to investigate the connections between *multi-amalgamation* [9] and recursive PINs [19].

Concerning dedicated tool support, we are currently working on integrating the approach in our model transformation tool eMoflon[5], which already decomposes complex patterns into PINs in order to be able to exploit the advantages of an underlying incremental pattern matcher [15]. Our formalisation and the flattening algorithm proposed in this paper can be useful for ensuring that our various decomposition and PIN optimisation strategies are semantics preserving.

Finally, we are working on a systematic evaluation of various incremental pattern matchers that accept PIN-based input with respect to the effect of network structure on scalability.

References

1. Anjorin, A., Buchmann, T., Westfechtel, B.: The families to persons case. In: TTC 2017, Marburg, Germany, 21 July 2017, pp. 27–34 (2017)
2. Bang, J.Y., et al.: CoDesign: a highly extensible collaborative software modeling framework. In: ICSE 2010, pp. 243–246. IEEE (2010)
3. Bergmann, G.: Incremental model queries in model-driven design. Ph.D. thesis, Budapest University of Technology and Economics, Budapest, October 2013. http://home.mit.bme.hu/bergmann/download/phd-thesis-bergmann.pdf. Accessed 16 Nov 2018
4. Bergmann, G., Horváth, Á., Ráth, I., Varró, D.: A benchmark evaluation of incremental pattern matching in graph transformation. In: Ehrig, H., Heckel, R., Rozenberg, G., Taentzer, G. (eds.) ICGT 2008. LNCS, vol. 5214, pp. 396–410. Springer, Heidelberg (2008). https://doi.org/10.1007/978-3-540-87405-8_27
5. Bergmann, G., Ökrös, A., Ráth, I., Varró, D., Varró, G.: Incremental pattern matching in the VIATRA model transformation system. In: GRaMoT 2008, pp. 25–32. ACM, New York (2008)
6. Beyhl, T., Blouin, D., Giese, H., Lambers, L.: On the operationalization of graph queries with generalized discrimination networks. In: Echahed, R., Minas, M. (eds.) ICGT 2016. LNCS, vol. 9761, pp. 170–186. Springer, Cham (2016). https://doi.org/10.1007/978-3-319-40530-8_11
7. Ehrig, H., Ehrig, K., Prange, U., Taentzer, G.: Fundamentals of Algebraic Graph Transformation. MTCSAES. Springer, Heidelberg (2006). https://doi.org/10.1007/3-540-31188-2

---

[5] www.emoflon.org.

8. Garzón, M.A., Lethbridge, T.C., Aljamaan, H., Badreddin, O.: Reverse engineering of object-oriented code into Umple using an incremental and rule-based approach. In: CASCON 2014, pp. 91–105. IBM Corp., Riverton (2014)
9. Golas, U., Ehrig, H., Habel, A.: Multi-amalgamation in adhesive categories. In: Ehrig, H., Rensink, A., Rozenberg, G., Schürr, A. (eds.) ICGT 2010. LNCS, vol. 6372, pp. 346–361. Springer, Heidelberg (2010). https://doi.org/10.1007/978-3-642-15928-2_23
10. Habel, A., Pennemann, K.H.: Correctness of high-level transformation systems relative to nested conditions. Math. Struct. Comput. Sci. **19**(2), 245–296 (2009)
11. Hanson, E.N., Bodagala, S., Chadaga, U.: Trigger condition testing and view maintenance using optimized discrimination networks. IEEE Trans. Knowl. Data Eng. **14**(2), 261–280 (2002)
12. Hébert, M.: λ-presentable morphisms, injectivity and (weak) factorization systems. Appl. Categ. Struct. **14**(4), 273–289 (2006). https://doi.org/10.1007/s10485-006-9024-9
13. Horváth, A., Bergmann, G., Ráth, I., Varró, D.: Experimental assessment of combining pattern matching strategies with VIATRA2. Int. J. Softw. Tools Technol. Transf. **12**(3–4), 211–230 (2010)
14. Kuschke, T., Mäder, P., Rempel, P.: Recommending auto-completions for software modeling activities. In: Moreira, A., Schätz, B., Gray, J., Vallecillo, A., Clarke, P. (eds.) MODELS 2013. LNCS, vol. 8107, pp. 170–186. Springer, Heidelberg (2013). https://doi.org/10.1007/978-3-642-41533-3_11
15. Leblebici, E., Anjorin, A., Fritsche, L., Varró, G., Schürr, A.: Leveraging incremental pattern matching techniques for model synchronisation. In: de Lara, J., Plump, D. (eds.) ICGT 2017. LNCS, vol. 10373, pp. 179–195. Springer, Cham (2017). https://doi.org/10.1007/978-3-319-61470-0_11
16. Leblebici, E., Anjorin, A., Schürr, A., Hildebrandt, S., Rieke, J., Greenyer, J.: A comparison of incremental triple graph grammar tools. In: ECEASST 67 (2014)
17. Varró, D., Bergmann, G., Hegedüs, Á., Horváth, Á., Ráth, I., Ujhelyi, Z.: Road to a reactive and incremental model transformation platform: three generations of the VIATRA framework. Softw. Syst. Model. **15**(3), 609–629 (2016)
18. Varró, G., Deckwerth, F.: A rete network construction algorithm for incremental pattern matching. In: Duddy, K., Kappel, G. (eds.) ICMT 2013. LNCS, vol. 7909, pp. 125–140. Springer, Heidelberg (2013). https://doi.org/10.1007/978-3-642-38883-5_13
19. Varró, G., Horváth, Á., Varró, D.: Recursive graph pattern matching. In: Schürr, A., Nagl, M., Zündorf, A. (eds.) AGTIVE 2007. LNCS, vol. 5088, pp. 456–470. Springer, Heidelberg (2008). https://doi.org/10.1007/978-3-540-89020-1_31
20. Wang, S., Morin, B., Roman, D., Berre, A.J.: A semi-automatic transformation approach for semantic interoperability. In: NATO Symposium and Workshop on Semantic & Domain Based Interoperability (2011)

Hybrid Search Plan Generation
for Generalized Graph Pattern Matching

Matthias Barkowsky[(✉)] and Holger Giese

Hasso-Plattner Institute, University of Potsdam,
Prof.-Dr.-Helmert-Str. 2-3, 14482 Potsdam, Germany
{matthias.barkowsky,holger.giese}@hpi.de

Abstract. In recent years, the increased interest in application areas such as social networks has resulted in a rising popularity of graph-based approaches for storing and processing large amounts of interconnected data. To extract useful information from the growing network structures, efficient querying techniques are required.

In this paper, we propose an approach for graph pattern matching that allows a uniform handling of arbitrary constraints over the query vertices. Our technique builds on a previously introduced matching algorithm, which takes concrete host graph information into account to dynamically adapt the employed search plan during query execution. The dynamic algorithm is combined with an existing static approach for search plan generation, resulting in a hybrid technique which we extend by a more sophisticated handling of filtering effects caused by constraint checks. We evaluate the presented concepts empirically based on an implementation for our graph pattern matching tool, the Story Diagram Interpreter, with queries and data provided by the LDBC Social Network Benchmark.

1 Introduction

In recent years, the increased interest in application areas such as social networks has resulted in a rising popularity of graph-based approaches for storing and processing large amounts of information [2]. The considered graphs frequently exhibit an inhomogeneous structure, which in the case of social networks can be caused by the diverse behavior of different users, including extreme outliers such as celebrities. In order to extract useful information from the growing, heterogeneous network structures, efficient querying techniques are required.

In this paper, we focus on queries without nesting and paths of varying length, which corresponds to the problem of graph pattern matching. Existing solutions usually work by iteratively mapping elements from a query specification to elements in a host graph according to a search plan. Since the order in which the

---

This work was developed mainly in the course of the project modular and incremental Global Model Management (GI 765/8-1), which is funded by the Deutsche Forschungsgemeinschaft.

E. Guerra and F. Orejas (Eds.): ICGT 2019, LNCS 11629, pp. 212–229, 2019.
https://doi.org/10.1007/978-3-030-23611-3_13

individual elements are mapped has a substantial impact on performance, many solutions employ sophisticated strategies for determining good search plans.

The majority of these techniques only considers structural information, that is typing information and edges between nodes in the graph like relationships between persons in a social network, for guiding the matching process. However, in many realistic application scenarios, nonstructural information also plays an important role. This includes attributes of nodes like the age of a person in the network as well as external data structures such as indices, which are particularly relevant in the context of graph databases or the evaluation of decomposed queries [6]. Hence, a tighter integration of constraints specified over such nonstructural information into the matching process is desirable.

We therefore introduce a unified notion of constraints in a graph query. We then propose a matching strategy that is based on an existing dynamic algorithm [12], which generates a search plan on the fly as a query is being executed. On the one hand, the dynamic technique allows tailoring the search to heterogeneities in the host graph that cannot be handled by a static search plan. On the other hand, this approach has the drawback of not being able to consider the overall structure of the query, which can lead to shortsighted decisions during the matching process. To address this problem, we combine our adapted approach with a static but model-sensitive technique for search plan generation [17]. The resulting hybrid approach is subsequently evaluated empirically using queries and datasets from the LDBC Social Network Benchmark [10].

The remainder of the paper is structured as follows: Sect. 2 briefly introduces the basic notion of graphs, graph morphisms, and graph queries as used in this paper. We then present our generalized approach for dynamic search plan generation in Sect. 3. In Sect. 4, we first integrate an existing static technique with our dynamic solution. The resulting hybrid approach is then extended to allow a more sophisticated consideration of constraint checks during search plan generation. The developed concepts are evaluated empirically in Sect. 5, using a benchmark from the domain of social networks. Section 6 discusses related work and Sect. 7 concludes the paper.

2   Prerequisites

We briefly reintroduce the notion of graphs and graph morphisms [9]. A *graph* $G = (G^V, G^E, s^G, t^G)$ consists of a set of vertices G^V, a set of edges G^E, a source function $s^G : G^E \to G^V$ and a target function $t^G : G^E \to G^V$. Given two graphs $G = (G^V, G^E, s^G, t^G)$ and $H = (H^V, H^E, s^H, t^H)$, a *graph morphism* $f : G \to H$ is a pair of mappings $f^V : G^V \to H^V$ and $f^E : G^E \to H^E$ such that $f^V \circ s^G = s^H \circ f^E$ and $f^V \circ t^G = t^H \circ f^E$. If f^V and f^E are injective, f is called a *monomorphism*.

A *typed graph* is a tuple $(G, type)$, where G is a graph and $type : G \to TG$ is a graph morphism into a *type graph* $TG = (TG^V, TG^E, s^{TG}, t^{TG})$. A *typed graph morphism* $f : G_1^T \to G_2^T$ between typed graphs $G_1^T = (G_1, type_1)$ and $G_2^T = (G_2, type_2)$ is a graph morphism $f : G_1 \to G_2$ such that $type_2 \circ f = type_1$.

A graph query is then specified by a typed *query graph* Q. A solution for a graph query given a typed *host graph* H is a typed monomorphism $m : Q \rightarrow H$ called *match*. Typically, explicit mappings are only computed for the query graph vertices and edges are mapped implicitly [3,7], which we adopt in this paper. The process of finding matches for a graph query in a given host graph is called graph pattern matching and corresponds to the execution of the query.

This usually involves executing a sequence of primitive search operations called *search plan*, which extends a partial, potentially empty monomorphism to a complete match. A naïve search plan may first create a mapping for each query vertex and subsequently check whether there is a corresponding host graph edge for each edge in the query graph. However, the required computational effort can vary substantially depending on the chosen plan. Hence the generation of efficient search plans is an important subtask of graph pattern matching.

In practice, the specification of a graph query may also contain additional constraints over the vertices of the query graph that further restrict the solution. This includes constraints over possible attributes or external data structures such as indices. Such information is called *nonstructural information* whereas the information encoded in the graph structure is considered *structural information*.

Example 1. Figure 1 shows an example graph query from a social network domain and the corresponding type graph, represented by a data model. The query is presented in Story Diagram notation [12], using an UML object diagram to denote the query graph Q and a box with shadow to indicate that all matches should be found. Q consists of two vertices of type *Person p1* and *p2*, one vertex of type *KnowsLink*, and two edges. The query also contains a constraint that

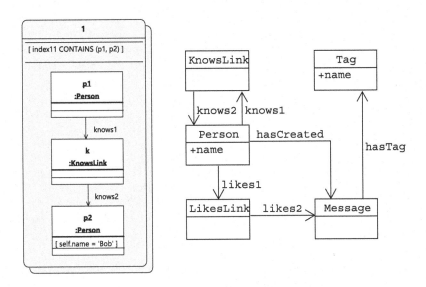

Fig. 1. Example graph query containing different kinds of constraints and the corresponding data model, which is an excerpt from the data model from [10]

requires some indexing structure *index11* to contain the tuple of mappings for
p1 and p2, and a constraint over the *name* attribute of the mapping for *p2*.

3   Search Model and Dynamic Search Plan Generation

In order to represent all possible kinds of constraints specified in a graph query in
a uniform manner, we developed a *Search Model* for graph queries. As displayed
in Fig. 2, a Search Model consists of three types of elements, some of which are
augmented with states to encode the state of a query execution.

Pattern Nodes represent vertices in the query graph Q and can either be
in state BOUND, indicating that a mapping for the Pattern Node has already
been determined, or UNBOUND otherwise. Pattern Nodes in state BOUND
also store the vertex in the host graph H which they are currently mapped
to. The *configuration* of a Search Model is given by the states of its Pattern
Nodes and hence encodes the domain of a partial graph morphism from the
encoded query graph into a host graph. The *state* of a Search Model comprises
its configuration as well as the *mapping* attributes of its Pattern Nodes and
therefore represents a partial graph morphism. We use C to denote the set of
possible configurations and S to denote the set of possible Search Model states.
The configuration of a Search Model state $s \in S$ is denoted by $config(s)$. The
configurations where all Pattern Nodes are in state UNBOUND respectively
BOUND are called the *empty configuration* $c_\emptyset$ and *complete configuration* c_C.
We define the *empty Search Model state* $s_\emptyset$ such that $config(s_\emptyset) = c_\emptyset$ and the
set of *complete Search Model states* $S_C := \{s_C | s_C \in S \wedge config(s_C) = c_C\}$. Each
$s_C \in S_C$ then represents a typed monomorphism $m : Q \to H$.

Constraints such as edges in the query graph or conditions over attributes
are represented by *Pattern Constraints*. Pattern Constraints may have a number
of dependencies that indicate over which Pattern Nodes the constraint is formulated. A Pattern Constraint in state ACTIVE has already been checked, whereas

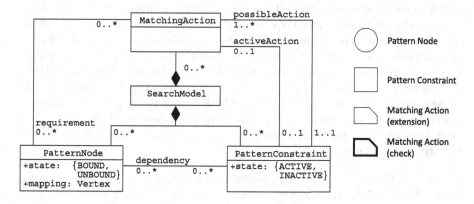

Fig. 2. Search Model metamodel and concrete syntax

an INACTIVE Pattern Constraint still needs to be handled. Additionally, Pattern Constraints are associated with *Matching Actions*, each of which encodes an executable operation that can be performed by a pattern matcher to ensure that the associated constraint is satisfied. Examples of Matching Actions include looking up host graph vertices of a certain type or checking the existence of an edge between two host graph vertices. For each Pattern Constraint, at most one Matching Action can be qualified as that constraint's *active action*, indicating that it is currently appropriate for handling the associated constraint.

Our approach allows arbitrary constraints to be translated into Pattern Constraints. In this paper, we focus on four exemplary kinds of constraints. A *domain constraint* restricts the type of a single vertex in the graph query. An *edge constraint* specifies that an edge of a certain type has to be present between a source and target query vertex. Both domain constraints and edge constraints are directly encoded in the query graph. *Index constraints* are formulated over an external data structure, requiring that it either contains or excludes some tuple. In this paper, we limit this to tuples of host vertices, which for example are useful for storing intermediate matches in the context of decomposed complex queries [6]. Lastly, we consider *expression constraints*, that is constraints formulated over attributes of query vertices in some expression language. Here, we restrict expression constraints to constraints over a single vertex.

Matching Actions may have *requirement* associations to Pattern Nodes. A Matching Action m is *applicable* to a Search Model state s with host graph H iff all of its requirements are in state BOUND in $c = config(s)$. The application of m then yields a set of states $S' \subseteq S$ denoted by $s[m\rangle_H S'$. The application of m to s always results in the same configuration ($\forall s' \in S' : config(s') = c'$) and thus for $config(s) = c$ we have the related application of m for configurations denoted by $c[m\rangle c'$. Each $s' \in S'$ thus encodes an extension of the graph morphism represented by s by mappings for Pattern Nodes that are UNBOUND in c and BOUND in c'. For a sequence of Matching Actions $w = m_1...m_n$ and configurations $c_0 c_1...c_n$ with $c_i[m_{i+1}\rangle c_{i+1}$ for $i \in [0, n-1]$, we write $c_0[w\rangle c_n$.

We distinguish two kinds of Matching Actions. A *check* m_e represents an actual checking of a constraint. It requires all dependencies of its Pattern Constraint and its execution does not change the current state, thus $s[m_e\rangle\{s\}$ iff the check is successful and $s[m_e\rangle\emptyset$ otherwise. An *extension* only has a subset of these dependencies as its requirements and extends the current mapping by mappings for all remaining dependencies such that the created overall mapping satisfies the associated constraint. The check of a Pattern Constraint corresponding to an edge constraint for instance checks whether an edge is present between a source and target host graph vertex. The traversal of host graph edges to collect candidate mappings for some Pattern Node on the other hand corresponds to an extension action of an edge constraint.

Example 2. Figure 3 shows an example graph query and the corresponding Search Model, using the concrete syntax from Fig. 2. Note that for readability, domain constraints are not visualized and Matching Actions are labeled with a set encoding their requirement links. The Search Model contains

Pattern Constraints corresponding to the edge constraints $e1$ and $e2$, the index constraint $i1$, and the expression constraint $c1$, as well as exemplary Matching Actions. These include edge lookup, traversal, reverse traversal and check for $e1$ and $e2$, index enumeration and check for $i1$ and the check of $c1$. Initially, the Search Model is in configuration $c_\emptyset$ and therefore state $s_\emptyset$. Because it requires no Pattern Nodes to be BOUND, the Matching Action $m1$ is applicable. Its application to $c_\emptyset$ then yields a configuration c_1 where Pattern Nodes $p1$ and $p2$ are in state BOUND. Its application to $s_\emptyset$ hence yields a set of states S_1 with configuration c_1 and all possible mappings for $p1$ and $p2$ that satisfy $i1$.

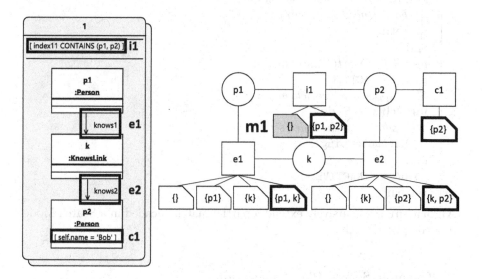

Fig. 3. Example query and corresponding Search Model

3.1 Search-Model-Based Matching Algorithm

In Algorithm 1, we outline our method for evaluating a graph query using a corresponding Search Model, which is a generalization of the dynamic matching algorithm introduced in [12]. The algorithm takes a Search Model state encoding a potentially empty partial match as an input and emits all complete matches that are extensions of that partial match.

The procedure starts with a check whether all of the Search Model's Pattern Constraints are in state ACTIVE. If the check is successful, the mapping currently encoded in the Search Model is a mapping for all vertices in the graph query that satisfies all specified constraints and can be emitted. Otherwise, one of the remaining Pattern Constraints in state INACTIVE that has an active action is selected, its state is set to ACTIVE and its active action is applied. The algorithm then loops over the generated set of extended Search Model states.

For each element, the Search Model is first updated with the candidate state. This includes updating the state and mapping of all affected Pattern Nodes and the active action of Pattern Constraints that have a dependency to an updated Pattern Node. Then a recursive call is performed to find all extensions of the updated Search Model state to complete states that satisfy all constraints. At the end of the procedure, the state of the selected Pattern Constraint is reset.

Procedure ExtendMapping(s)

> **Input** : s: A Search Model state encoding a partial match
> **Output**: All complete extensions of the encoded partial match
>
> **if** $AllConstraintsActive()$ **then**
> > $EmitEncodedMapping(s)$;
> > **return**;
>
> **end**
> $p \leftarrow SelectNextConstraint(s)$;
> $p.state \leftarrow$ ACTIVE;
> $S' \leftarrow apply(p.activeAction, s)$;
> **foreach** $s' \in S'$ **do**
> > $UpdateSearchModel(s')$;
> > ExtendMapping(s');
>
> **end**
> $p.state \leftarrow$ INACTIVE;
> **return**;

Algorithm 1: Recursively extend a partial match encoded in a Search Model to a complete match

3.2 Dynamic Search Plan Generation

The algorithm presented in Sect. 3.1 provides a degree of freedom with respect to the selection of the next Pattern Constraint to check, which corresponds to a single step in a dynamically generated search plan for the query. Which Pattern Constraint with active action m is chosen in each step has a significant impact on the remaining effort, which can be defined recursively for the current Search Model state s and host graph H for S' uniquely defined by $s[m\rangle_H S'$ as

$$cost_S(s, H) := \sum_{s' \in S'} (1 + cost_S(s', H)) = \sum_{s' \in S'} 1 + \sum_{s' \in S'} cost_S(s', H), \quad (1)$$

with $cost_S(s_C, H) := 0$ for $s_C \in S_C$.

Note that we exclude the cost for updating the Search Model, which is a rather inexpensive operation for small queries over large host graphs [5] as confirmed by our experimental results in Sect. 5. The overall effort is then given by $cost_S(s_\emptyset, H)$. However, as this effort depends on the concrete recursive search through H, it cannot directly be employed to choose the next Pattern Constraint.

Instead, we compute the cost of a Matching Action m in s as the growth of the search space resulting from its application by considering only the first part $\sum_{s' \in S'} 1$ of Eq. 1 for S' uniquely defined by $s[m\rangle_H S'$ using

$$cost_M^{dyn}(s, m, H) := \sum_{s' \in S'} 1 = |S'|. \tag{2}$$

Similar to the approach in [12], the matching algorithm can then choose the next Pattern Constraint p in a greedy manner, such that the related Matching Action $m = p.activeAction$ has minimal cost $cost_M^{dyn}(s, m, H)$. As an exception, we execute any applicable check Matching Action as soon as possible according to the fail-first principle [13].

4 Hybrid Search Plan Generation

The fully dynamic strategy for search plan generation described in Sect. 3.2 is often able to determine a good search plan if the Search Model only contains few Matching Actions with a potentially large candidate set size. However, it disregards the overall structure of the query. In some cases, it may therefore select Pattern Constraints with Matching Actions that seem appealing in the context of the current candidate mapping, but are a suboptimal choice in the context of the remainder of the query. This problem is addressed by matching strategies that precompute a static search plan based on statistics collected from the host graph, such as the average number of outgoing edges of a vertex. These approaches however do not account for heterogeneities in the host graph where for example a vertex has a significantly above average outdegree.

4.1 Combining Static and Dynamic Search Plan Generation

To leverage the more accurate information available during the execution of a graph query while still considering the overall structure of the query, we propose a hybrid strategy for search plan generation. The combined approach can easily be integrated with the Search-Model-based matching algorithm presented in Sect. 3.1 in the form of a strategy for Pattern Constraint selection and is based on an adapted cost function for Matching Actions. The adapted cost function no longer only considers the growth of the search space in each step as given by Eq. 2. It rather computes an estimate for the size of the search space of the entire remainder of the query. This computation is based on (i) the size of the Matching Action's candidate set and (ii) a precomputed estimate for the size of the search space left after selecting one of the candidate mappings. The former corresponds to the cost function for Matching Actions from Sect. 3.2. To obtain the latter, we build on a static technique by Varró et al. [17].

Varró et al. use a dynamic programming algorithm to gradually generate search plans that lead from an initial state to each reachable search state. A search state in [17] corresponds to a Search Model configuration, whereas a search plan corresponds to a sequence of extension Matching Actions $w = m_1...m_l$ with a sequence of configurations $c_0 c_1 ... c_l$ such that $\forall i \in [1, l] : c_{i-1}[m_i)c_i$. Each plan is associated with a static estimate for the overall effort from Eq. 1, given by $cost_C^{sta}(c_i, H, m_1 ... m_l) := cost_M^{sta}(m_i, H) * (1 + cost_C^{sta}(c_{i+1}, H, m_1 ... m_l))$ and $cost_C^{sta}(c_l, H, m_1 ... m_l) := cost_M^{sta}(m_l, H)$, where $cost_M^{sta}(m_i, H)$ is an estimate for the cost of a Matching Action ($|S'|$) based on statistical data collected from the host graph and the size of considered indices. This can be summarized for the complete search plan with $c_0 = c_\emptyset$ and $c_l = c_C$ to

$$cost_C^{sta}(c_\emptyset, H, m_1 ... m_l) := \sum_{j=1}^{l} \prod_{i=1}^{j} cost_M^{sta}(m_i, H). \tag{3}$$

We can generalize Eq. 3 to $cost_C^{sta}(c, H, w)$ for an arbitrary configuration c and search plan w with $c[w)c'$ for target configuration c' and thus obtain the minimal cost for that transition by

$$cost_{CC}^{sta}(c, c', H) := \min_{w \text{ with } c[w)c'} cost_C^{sta}(c, H, w). \tag{4}$$

As a byproduct of finding the optimal search plan that leads from $c_\emptyset$ to c_C, the dynamic programming algorithm by Varró et al. computes $cost_{CC}^{sta}(c_\emptyset, c, H)$ for all reachable $c \in C$. However, for our hybrid approach we require the values for $cost_{CC}^{sta}(c, c_C, H)$ for all configurations $c \in C$ to the complete configuration c_C.

Therefore, rather than performing search plan generation starting from $c_\emptyset$ as in the approach by Varró et al., we compute search plans and their associated cost backwards starting with c_C. Intuitively, we use the dynamic programming approach to undo Matching Action applications, creating configurations and search plans that lead from these configurations to c_C. Thus, we obtain a table containing the best generated partial search plan starting in each configuration, as well as the associated cost.

Based on $cost_M^{dyn}$ and $cost_{CC}^{sta}$, we construct a new, hybrid cost estimation function for extension Matching Action m with configuration c' uniquely defined by $c = config(s)$ and $c[m)c'$

$$cost_M^{hyb}(s, m, H) := cost_M^{dyn}(s, m, H) * (1 + cost_{CC}^{sta}(c', c_C, H)). \tag{5}$$

Note that to compute an optimal search plan with respect to $cost_{CC}^{sta}$, the approach in [17] requires exponential effort in the size of the query. This is also true for our hybrid technique, since it has to determine search plans for all possible configurations. However, the table only has to be computed once before the execution of the query in both cases. Furthermore, in many application areas query graphs tend to be rather small, but are matched into large host graphs. In these cases the actual execution time dominates the time for the preliminary computations as demonstrated in Sect. 5.

Example 3. Consider the example query from [10] shown in Fig. 4. Assuming the pattern matching has already determined a mapping for query vertex *p2*, there are two viable search plans for the remainder of the query, either following the edges in clockwise or counterclockwise direction. A fully dynamic approach runs the risk of choosing neither and enumerating a cartesian product of mappings for *p1* and *m*. An adequate static approach would decide for one plan and adhere to it for every instance of *Person*. However, if the number of incoming *knows2* and outgoing *likes1* associations varies for different mappings for *p2*, the search plan with minimal actual effort as given by Eq. 1 may vary as well. The introduced hybrid approach is able to switch between the plans dynamically while avoiding the cartesian product.

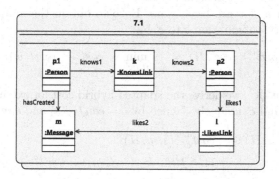

Fig. 4. Query used for evaluation in Sect. 5 with multiple viable search plans

4.2   Hybrid Search Plan Generation with Filtering Effects

So far, the approach presented in [17] and our related hybrid approach only consider extension actions in the cost function for search plans. Explicit check actions are simply inserted into the search plan as soon as possible, without effect on the overall cost. However, performing a check potentially cuts off part of the remaining search space if the check fails. It is therefore useful to also consider checks in the cost function to steer search plan generation. Hence, similarly to the cost associated with extension Matching Actions, we associate Pattern Constraints p with a filtering rate f_p in $[0, 1]$. The filtering rate represents an estimate of the portion of possible candidate mappings for the constraint's dependencies satisfying the constraint. It is thus also an estimate for the portion of the search space remaining after cutting off the parts that fail the check of that constraint. We define the combined filtering rate of a set of Pattern Constraints P as $flt(P) := \prod_{p \in P} f_p$.

For a configuration c, we denote the set of Pattern Constraints for which all dependencies are BOUND in c by $chk_C(c)$. For configurations c and c' and an extension Matching Action m with $c[m\rangle c'$, the set of Pattern Constraints for which a check can be executed after applying m to c but not before is then given by $chk(c, c', m) := chk_C(c') \setminus (chk_C(c) \cup \{m.constraint\})$.

To also consider the filtering rates, we adapt the cost function $cost_C^{sta}$ for configuration c, host graph H and a search plan consisting of a sequence of extension actions $w = m_1 \dots m_l$ with configurations $c_0 \dots c_l$ such that $c_0 = c$, $c_i[m_i\rangle c_{i+1}$ for $i \in [0, l-1]$, and $c_l = c_C$ as follows

$$cost_C^{sta,f}(c, H, m_1..m_l) := \sum_{j=1}^{l} \prod_{k=1}^{j-1} flt(chk(c_{k-1}, c_k, m_k)) \prod_{i=1}^{j} cost_M^{sta}(m_i, H), \quad (6)$$

as for each Matching Action, the filtering of all Pattern Constraints for which checks were executed beforehand matters.

We can then as for Eq. 4 generalize Eq. 6 to obtain the minimal cost for a transition from a configuration c to c' by

$$cost_{CC}^{sta,f}(c, c', H) := \min_{w \text{ with } c[w\rangle c'} cost_C^{sta,f}(c, H, w). \quad (7)$$

Based on $cost_{CC}^{sta,f}$, we derive the adapted hybrid cost for extension Matching Action m for c and c' uniquely defined by $c = config(s)$ and $c[m\rangle c'$ by

$$\begin{aligned} cost_M^{hyb,f}(s, m, H) := \; & cost_M^{dyn}(s, m, H) * \\ & (1 + flt(chk_M(c, c', m)) * cost_{CC}^{sta,f}(c', c_C, H)). \end{aligned} \quad (8)$$

We use the statistical data collected for estimating the cost of extension actions and index sizes to compute estimates for the filtering rate of Pattern Constraints. However, in addition to the data collected in [17], for each expression constraint with dependency n, we check for each host graph vertex that matches the type of the query vertex represented by n whether it satisfies the expression. We store the number of successful checks to compute the filtering rate for the constraint's check Matching Action.

Example 4. Figure 5 shows another example query from [10] where filtering of explicit constraint checks can have a substantial performance impact. If the

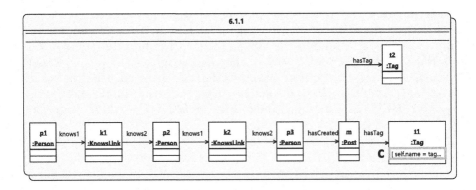

Fig. 5. Query used for evaluation in Sect. 5 with a check acting as a filter

expression constraint c over query vertex $t1$ is only satisfied by a small number of $Tags$, it is highly preferable to start the matching process there to rule out large portions of the search space as early as possible. This is considered by our extended cost function.

5   Evaluation

We implemented the presented concepts in our EMF-based Story Diagram Interpreter tool [12]. Since backwards traversal of edges and edge lookup are not generally supported by EMF [1], we extended our tool to allow the corresponding Matching Actions. For performance evaluation of simple queries without nesting and paths of varying length, we decomposed twelve of the 14 complex reading queries of the Interactive Workload of the LDBC Social Network Benchmark [10] according to their formulation in natural language. We omitted the two remaining queries as they include paths of arbitrary length. For queries with paths of varying but bounded length, we chose the maximum length. Similar to our approach in [6], we encoded nesting as index constraints. This yielded a total of 32 partial queries corresponding to the task of matching simple graph patterns.

We then executed Story Diagram implementations of the partial queries over five datasets of different size generated using the data generator of the benchmark, which we parametrized with different numbers of persons in the network (500, 1000, 1500, 2000, 2500), obtaining host graphs with up to 1 500 000 vertices and 4 300 000 edges. For search plan generation, we considered three strategies:

- **DYNAMIC**: fully dynamic approach as introduced in Sect. 3.2
- **STATIC**: static approach presented by Varró et al. [17]
- **HYBRID**: hybrid approach as introduced in Sect. 4.2

All experiments were performed on a Linux SMP Debian 4.9.18-1 machine with Intel Xeon E5-2640 CPU (2.5 GHz clock rate) and 32 GB main memory running OpenJDK version 1.8.0_181. All presented times correspond to the average time measured in five runs of the respective experiment, each including a brief warm-up. For all experiments with an average execution time greater than 1 s, the standard deviation was less than 10% of the average time.

5.1   Experimental Results

Figure 6 shows the measured execution times. Note that for readability, we included only those partial queries for which the performance of HYBRID was better or worse than the performance of either reference strategy by at least factor 1.5 for at least one dataset and the execution took more than 1 s. DYNAMIC outperforms STATIC for query 5.1.1, whereas STATIC achieves better results than DYNAMIC for queries 8.1 and 10.1 and has a more stable performance for query 11.1. The results of HYBRID are always at least close to the better performing reference strategies and are significantly better in some cases.

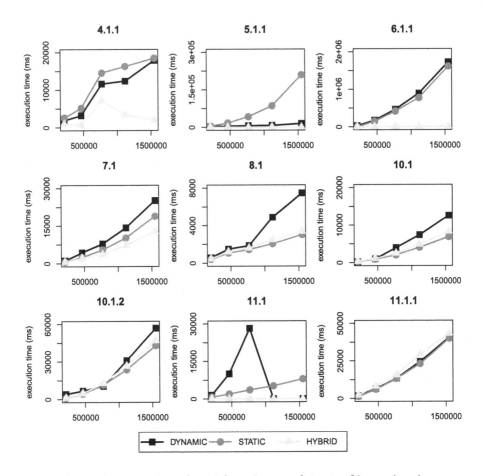

Fig. 6. Execution time of partial queries over datasets of increasing size

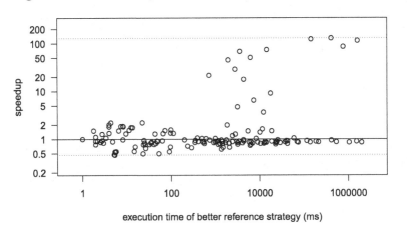

Fig. 7. Speedup of HYBRID compared to the better reference strategy

Considering all 32 partial queries and all datasets, the execution time of HYBRID is at most 2.1 times the execution time of the better reference strategy (speedup 0.48), as shown in Fig. 7. At the same time, the HYBRID strategy speeds up the execution of some queries by up to factor 100. The slight decrease in performance for some cases can be explained by a small overhead associated with the updating of the Search Model when comparing to the STATIC technique and the time required to compute the cost tables when comparing to DYNAMIC. The overhead is particularly pronounced for queries with a low execution time.

For queries 4.1.1, 5.1.1, 6.1.1 and 11.1, the performance gain of HYBRID stems from the improved handling of filtering effects. These queries include expression constraints, which are ignored by the DYNAMIC and STATIC strategy until a mapping for the respective vertex is created. HYBRID considers these constraints during search plan generation. In the case of query 6.1.1, which was introduced in Example 4, this results in HYBRID matching the vertex $t1$ first. This significantly cuts the search space compared to DYNAMIC and STATIC, which both start with matching $p1$. The advantage of the hybrid generation of search plans is visible for query 7.1, which was presented in Example 3. For further evaluation, we executed this query over the largest dataset for each *Person* vertex in the network, which we used as an initial mapping for $p2$. For each of these roughly 2200 individual executions, we measured the size of the traversed search space by counting the total number of candidate mappings created for all query vertices. Figure 8 shows a histogram of the measurement results.

Figure 8 shows that the DYNAMIC strategy often achieves very small search spaces, but produces outlier performances by sometimes employing an inadequate search plan that results in the enumeration of a large cartesian product. This is illustrated by a number of mappings leading to very large search spaces. The STATIC strategy adheres to a single, precomputed search plan for all mappings of $p2$, which results in a more stable performance. However, it misses out on the opportunity to reduce search space size via a different search plan and hence rarely achieves very small search spaces. By also considering concrete host graph information, the HYBRID strategy is able to dynamically switch search plans while using the static information to avoid plans that lead to cartesian products. As a result, it creates fewer than 1000 candidate mappings for about 40% of the initial mappings and only creates more than 10 000 for less than 10%.

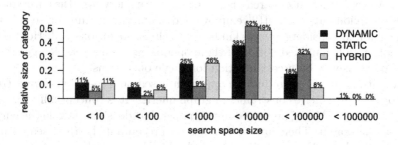

Fig. 8. Number of mappings for $p2$ leading to certain search spaces for query 7.1

The collection of statistical data required for the preliminary computations of the STATIC and HYBRID strategy can mostly be integrated with the construction of the indexing structures required for backwards traversal and lookup of edges, causing no further overhead compared to the DYNAMIC strategy. Also, these indices only have to be created once per host graph. The filtering rates of expression constraints are an exception, since they are not required for the DYNAMIC or STATIC strategy and have to be computed for each individual constraint of a query that is executed for the first time. The time to compute the filtering rates was below 500 ms for all partial queries. This seems feasible for many partial queries where the overall execution time was much higher.

To obtain optimal static search plans, the computation of the search plan tables for HYBRID and STATIC is required whenever a query is executed over a changed dataset. This is potentially problematic since the computation has an exponential runtime complexity. However, for our examples the required time was always below 50 ms for both strategies and within 1.5 times of each other.

Our evaluation demonstrates that our technique is competitive and can outperform existing approaches for a realistic application scenario as represented by the independent and widely accepted LDBC Social Network Benchmark. Regarding threats to validity, we remark that the results are not necessarily generalizable to different domains, which would require further experiments.

6   Related Work

A variety of model transformation tools each employ their own heuristics for matching the left hand side of transformation rules, which corresponds to the execution of a graph query. PROGRES [18] considers several kinds of matching operations and offers a uniform representation in a *Pattern Graph*, but only employs a simple greedy strategy for search plan generation. GrGen [11] builds a *Plan Graph* for a query, which encodes matching actions related to domain and edge constraints. A search plan is then determined by computing a minimum spanning tree of the Plan Graph. In Henshin [3], the matching order is mainly determined based on domain constraints, but the filtering effect of attribute constraints is considered to some extent. The local search mechanism of EMF-IncQuery [8] employs the technique for search plan generation presented in [17], which we use as the foundation of our hybrid matching approach. All mentioned tools however use a static search plan to execute the query over the entire model.

Our previous approach [12] employs a dynamic algorithm for search plan generation and considers several kinds of matching operations [14]. However, it does not provide a unified notion of these operations and search plan generation is based exclusively on a cost function for single operations.

Horváth et al. [15] introduce an approach based on a so-called *Search Graph* for representing generic constraints in graph queries. It is conceptually very similar to the Search Model presented in this paper, but does not encode the current state of the search. Their approach relies on precomputed, static search plans and specifically addresses the timing of checks. However, they do not consider filtering effects but focus on the cost of performing the check.

In [7], a matching technique is proposed that is based on decomposing the query into a core, forest and leaf structure. The core structure comprises the dense part of a query and is matched first to exploit the filtering of edge constraints. Since the approach focuses on the pure subgraph matching problem, it does not consider constraints other than edge and domain constraints.

Bak and Plump design an efficient algorithm for graph pattern matching [4] based on designating certain vertices as *roots*. Roots are then used as starting points for the search, thus restricting the search space. While the notion of roots can be integrated with our approach in the form of Pattern Constraints and Matching Actions, it is not appropriate in every application scenario.

Incremental graph pattern matching as implemented in VIATRA [16] offers an alternative to local search. VIATRA stores all matches of a query in a cache, which is updated when the host graph changes. The updating process is realized via RETE networks, the structure of which corresponds to a search plan. This enables an efficient enumeration of matches when a query is repeatedly executed over an evolving host graph, but causes overhead in memory consumption.

7 Conclusion

In this paper, we developed an approach for graph pattern matching based on a Search Model representation of a graph query and a generalized version of the algorithm from [12]. We then integrated an existing static technique [17] with our dynamic algorithm and extended the resulting hybrid strategy by a more sophisticated handling of filtering effects. Finally, we empirically evaluated our solution using our Story Diagram Interpreter tool with decomposed queries and generated data from the LDBC Social Network Benchmark. Our results confirm that considering static information significantly improves the performance of the hybrid strategy compared to a fully dynamic approach. While the hybrid technique comes with an overhead in execution time compared to a static approach, its dynamic nature can improve performance for heterogenous host graphs. Our results also demonstrate the benefit of our improved handling of filtering effects.

As future work, we will study how the optimizations concerning the exponential execution time of table calculation introduced in [17] can be integrated with our hybrid matching technique and how the overhead for the required statistics can be decreased at the expense of accuracy, for instance by computing filtering rates based on a sample of host graph vertices.

We also plan to investigate how to take further advantage of the dynamic nature of our approach by integrating information about the concrete instance situation into the precomputed search plan tables. Finally, we will extend our evaluation by applying our technique to different application areas and compare our realization to other existing tools.

References

1. EMF: Eclipse Modeling Framework. https://www.eclipse.org/modeling/emf/. Accessed 7 May 2019
2. Angles, R.: A comparison of current graph database models. In: 2012 IEEE 28th International Conference on Data Engineering Workshops, pp. 171–177. IEEE (2012). https://doi.org/10.1109/ICDEW.2012.31
3. Arendt, T., Biermann, E., Jurack, S., Krause, C., Taentzer, G.: Henshin: advanced concepts and tools for in-place EMF model transformations. In: Petriu, D.C., Rouquette, N., Haugen, Ø. (eds.) MODELS 2010. LNCS, vol. 6394, pp. 121–135. Springer, Heidelberg (2010). https://doi.org/10.1007/978-3-642-16145-2_9
4. Bak, C., Plump, D.: Rooted graph programs. In: Proceedings of International Workshop on Graph-Based Tools (GraBaTs 2012), vol. 54 (2012). https://doi.org/10.14279/tuj.eceasst.54.780
5. Barkowsky, M.: Tight integration of indices into graph query execution. Master's thesis, Hasso Plattner Institute for Digital Engineering (2018)
6. Beyhl, T., Blouin, D., Giese, H., Lambers, L.: On the operationalization of graph queries with generalized discrimination networks. In: Echahed, R., Minas, M. (eds.) ICGT 2016. LNCS, vol. 9761, pp. 170–186. Springer, Cham (2016). https://doi.org/10.1007/978-3-319-40530-8_11
7. Bi, F., Chang, L., Lin, X., Qin, L., Zhang, W.: Efficient subgraph matching by postponing Cartesian products. In: Proceedings of the 2016 International Conference on Management of Data, pp. 1199–1214. ACM (2016). https://doi.org/10.1145/2882903.2915236
8. Búr, M., Ujhelyi, Z., Horváth, Á., Varró, D.: Local search-based pattern matching features in EMF-IncQuery. In: Parisi-Presicce, F., Westfechtel, B. (eds.) ICGT 2015. LNCS, vol. 9151, pp. 275–282. Springer, Cham (2015). https://doi.org/10.1007/978-3-319-21145-9_18
9. Ehrig, H., Ehrig, K., Prange, U., Taentzer, G.: Fundamentals of Algebraic Graph Transformation. MTCSAES. Springer, Heidelberg (2006). https://doi.org/10.1007/3-540-31188-2
10. Erling, O., et al.: The LDBC social network benchmark: interactive workload. In: Proceedings of the 2015 ACM SIGMOD International Conference on Management of Data, pp. 619–630. ACM (2015). https://doi.org/10.1145/2723372.2742786
11. Geiß, R., Batz, G.V., Grund, D., Hack, S., Szalkowski, A.: GrGen: a fast SPO-based graph rewriting tool. In: Corradini, A., Ehrig, H., Montanari, U., Ribeiro, L., Rozenberg, G. (eds.) ICGT 2006. LNCS, vol. 4178, pp. 383–397. Springer, Heidelberg (2006). https://doi.org/10.1007/11841883_27
12. Giese, H., Hildebrandt, S., Seibel, A.: Improved flexibility and scalability by interpreting story diagrams. Electron. Commun. EASST **18** (2009). https://doi.org/10.14279/tuj.eceasst.18.268
13. Haralick, R.M., Elliott, G.L.: Increasing tree search efficiency for constraint satisfaction problems. Artif. Intell. **14**(3), 263–313 (1980). https://doi.org/10.1016/0004-3702(80)90051-X
14. Hildebrandt, S.: On the performance and conformance of triple graph grammar implementations. Ph.D. thesis, Hasso Plattner Institute at the University of Potsdam, June 2014
15. Horváth, Á., Varró, G., Varró, D.: Generic search plans for matching advanced graph patterns. Electron. Commun. EASST **6** (2007). https://doi.org/10.14279/tuj.eceasst.6.49

16. Varró, D., Bergmann, G., Hegedüs, Á., Horváth, Á., Ráth, I., Ujhelyi, Z.: Road to a reactive and incremental model transformation platform: three generations of the VIATRA framework. Softw. Syst. Model. **15**(3), 609–629 (2016). https://doi.org/10.1007/s10270-016-0530-4

17. Varró, G., Deckwerth, F., Wieber, M., Schürr, A.: An algorithm for generating model-sensitive search plans for EMF models. In: Hu, Z., de Lara, J. (eds.) ICMT 2012. LNCS, vol. 7307, pp. 224–239. Springer, Heidelberg (2012). https://doi.org/10.1007/978-3-642-30476-7_15

18. Zündorf, A.: Graph pattern matching in PROGRES. In: Cuny, J., Ehrig, H., Engels, G., Rozenberg, G. (eds.) Graph Grammars 1994. LNCS, vol. 1073, pp. 454–468. Springer, Heidelberg (1996). https://doi.org/10.1007/3-540-61228-9_105

Author Index

Printed in the United States
By Bookmasters